HAYES ON RECORD

Edited by

Peter Hall and Colin Brown

ERRATUM HAYES ON RECORD

Facing
Page 1

    Caption   (see page 216)
    Should read   (see page 206)

Page 11

Photograph shows Laurie Bambe in No 2 Studio at Abbey Road, not, as suggested, at Head Office, Blyth Road

Page 12

Ever since Edison discovered that sound had a distinctive wave form,

Should read:
Ever since <u>Leon Scott</u> discovered

Page 63

Line 10
Four 7" presses

Should read:
Four 10" presses

Page 82

Valve 5; cylinder should be cylinders
Valve 6; cylinders should be cylinder

Page 163

Photograph shows Ted Peverall not
Tony Keeler

FOREWORD

When I had the opportunity to return to work at the EMI Records Hayes factory in 1988, I was delighted. Having worked there previously [from 1977 to 1981] I had some idea of what awaited me.

The thing I looked forward to most on my return was renewing some old acquaintances, and to working with what I consider one of the most loyal and friendly groups of people one could wish to meet. EMI had always encouraged an atmosphere in which individuals were able to be creative in the truest sense of the word, and, despite the problems that led to the take-over by THORN in the late 1970s, this attitude of experimentation had continued.

Nowhere was this inclination to try to develop methods and machinery more noticeable than in the disc factory. The move from 78s to microgroove discs, the development of automated machinery, and the changes caused by the massive expansion of sales in the 1960s and early 1970s, had all been tackled with vigour. The introduction of the compact disc, however, finally seems to have sounded the death-knell of the factory although even now new ideas are being introduced, and improvements to machinery continue to be made.

With the present decline in the demand for vinyl we have decided to try to record some of the history of the Hayes factories, both Blyth Road and Uxbridge Road, to give an idea of how the sites, the product and the manufacturing process developed, before the inevitable death of vinyl records causes the closure of the Hayes facility. This book has been written by various employees from their own experiences and what information they could find. The writers have attempted to cover the development of both the machinery and the methods used in the factory to produce records, and, later, reel-to-reel tapes and music cassettes. Some of this is fairly technical, and while we apologise for this, factories are the workplace of engineers and it is difficult to describe the process non-technically. Wherever possible we have included pictures and diagrams to assist the non-technical reader.

Sadly the thing that always impressed me most about the company, its staff, was the one area where historical information was extremely short, and memories could not recall many details of staff prior to the last war. At its peak the manufacturing operation was employing over 1,000 staff. This book is dedicated to the many thousands who, over the years, have played a part in making the factory the very special place it was, and ensuring that it stayed a friendly and happy environment within which to work.

As I write, the factory is entering probably its last days; it now employs fewer than 150 staff, most of whom are involved in making music cassettes. It continues to retain its spirit and atmosphere, despite the decline in staff in the last four years, to the present number from around 700 in late 1988. The vinyl record seems to be disappearing from daily life, as a natural part of technological progress, as did the shellac 78s. This decline in demand came upon us at a much faster rate than we anticipated: as late as November, 1990 the factory produced some 6,700,000 vinyl records in one month; now, however, output has declined to only 900,000 per month [June 1992].

EMI does, of course, manufacture compact discs in the U.K., but not in the Hayes area. So, unlike the move from 78s to microgroove records, where the factory changed from making the old product to making its replacement, it is likely that soon EMI will not be making music products in Hayes for the first time since 1908.

It is very difficult to convey how much this factory meant to us all. I can only hope that the reader will be able to distil this information from the book itself, and forgive us our nostalgic and occasionally rose-tinted view of what it was like here. There were difficult and hard times but, as with childhood, memories tend to recall only the sunny days, and there seemed a lot more of those at the Hayes factories than most places.

Peter Hall
December, 1992

ACKNOWLEDGEMENTS

Many thanks are due to those who have written the articles, Colin Brown, Chris Pooley, Jim Hughes, Robin Allen, Mike Russell, Terry Conroy, Beverley Harris, Jim Wilsher and Peter Hall.

Thanks also must go to those whose memories we have plundered, including Wally Rand, John Yeoman, John Simmons, Ron Turvey, John Combes, Bill Thorpe, Johnny Harris, Peter Kenny, Dennis Bendall, Jim Adams, Bill Soby and many others too numerous to mention.

Special thanks to Bernie Gould for rediscovering the lay-out drawing of the Blyth Road site. Credit should also be given to Dave Bowring and Peter Williams for transcribing this, and other drawings, into a form suitable for this book. Much appreciation also to Margo Holman for her sterling efforts in compiling the roll of honour, and typing for those writers unable to do so!

Special mention should be made of Colin Brown and Peter Hall who have undertaken the task of compiling and editing the book. Kate Dunning and Howard Tweedie are thanked for their help and advice. Grateful thanks also go to Ruth Edge and her team in the EMI Records Archives for their help not only with our research, but for making their incredible volume of historical material available to us. We look forward to the eventual establishment of her museum, an event we think long overdue! We hope that some small working part of the history of the factory may be included in it, so that posterity can enjoy the sight of EMI's record presses in action.

We apologise unreservedly to anyone who has not been mentioned and should have been, and to anyone whose contribution to the company or the book has been omitted. We can only claim that we have done our best to record our version of the history of the factory, both for those who worked there, and any others who may be interested.

INTRODUCTION

This book has been structured to provide an idea of the development of the factory, its machines and processes. It consists of five chapters, covering the sites, the equipment, the materials and the people.

Chapter one 'The Sites' begins with a general description of Hayes and its' transport and communication links. It then proceeds to a tour of the Blyth Road site, exploring the various buildings and their purposes. This is followed by a section describing a visitors' tour of the record factory, undertaken towards the late 1950s, a period when both 78s and microgroove records were being made. The tour does not cover the technical aspects of the operation in great detail, but is intended to provide an insight into the type of operations involved in manufacturing at that time.

The next part of this chapter describes the move from Blyth Road to Uxbridge Road, why it came about, and how it was achieved. It also looks at the new site, as it was first constructed. The next section describes a tour undertaken of this plant in 1990, and, once again, is non-technical in nature.

The final section of this chapter is concerned with EMI's other record factories. This has been included for the simple reason that many of these were designed and built by Hayes staff, and equipped with Hayes manufactured equipment. They were regarded, in many cases, as "children of Hayes". Indeed, Hayes became known as the "mother plant", because it provided most of them with full technical support throughout their lives.

Chapter two focuses more clearly on the development of the equipment used over the years to manufacture records. It covers the matrix area, the pressroom, quality control, packing and other equipment, as well as the plant services area. Also included is a section on the development of music cassette manufacturing. Although this operation uses almost entirely bought-in equipment, it has developed greatly during the product's lifetime, and is an integral part of the factory operation.

The next chapter explores the development of the materials used to manufacture records, and covers three basic areas. These are the record compound itself, from shellac to vinyl, the record labels, and the jackets and bags used.

The fourth chapter is concerned with the people and events in the plant. The first section covers wartime events at Blyth Road, including wartime experiences. The second is about the various sports and social activities at the sites, and is followed by sections on entertaining the workforce and visitors to the site. There is a section on the T.Q.M. programme launched at Uxbridge Road towards the end of the factory's life, which can be seen as an example of the type of relationship that existed between the staff and the managers. It was one generally based upon friendship and a sense of camaraderie. It focused on maintaining pride in the workplace, pride in its achievements and pride in its skills. Most importantly, it enabled people to feel proud when they said "I work at EMI's record factory". The remaining section contains brief descriptions of the men who have been "in charge" of the factory.

During the course of researching this book many people came forward with anecdotes concerning the factory and its staff. While we were not sure that these should be included, we felt that some of them were too good to miss out. Some had to be omitted, however, on the grounds of either good taste, or the libel laws!

The anecdotes have been included in the final chapter with the Roll of Honour of those who have completed 20 or more years service since Uxbridge Road opened. Unfortunately our records do not go back beyond then, and even some of the records we do have are not completely reliable.

The close relationship that existed in the development of the equipment, processes and materials has necessitated some repetition within the text. This has been done in order that the non-technical reader may more fully understand the way in which progress in one area creates the need for changes in another.

CONTENTS

Chapter One: The Factory Sites   1
    The Hayes Area and Its Communication and Transport Links
    The Blyth Road Site
    A Tour of The Blyth Road Record Factory
    The New Site
    A Tour of the Uxbridge Road Factory
    Overseas Factories

Chapter Two: The Equipment   69
    Introduction
    The Matrix Plant
    The Record Press
    Automation of the 7" Press
    The SP130 Automatic 7" Press
    Automation of the 12" Press
    The Type 1400 12" Press
    Music Test Equipment
    Packing Equipment
    Other Equipment
    Cassette Equipment
    The Power House

Chapter Three: The Materials   119
    Introduction
    Record Compounds
    Record Labels
    Record Packaging

Chapter Four: Working in the Factories   142
    Blyth Road During the War
    Sports and Social Activities
    Staff Sales
    Entertaining the Workforce
    Visitors to the Sites
    Health and Safety at Work
    Total Quality Management, Achievement Through People
    At the Helm, Manufacturing Managers

Chapter Five: 183
    Anecdotes
    The Roll of Honour

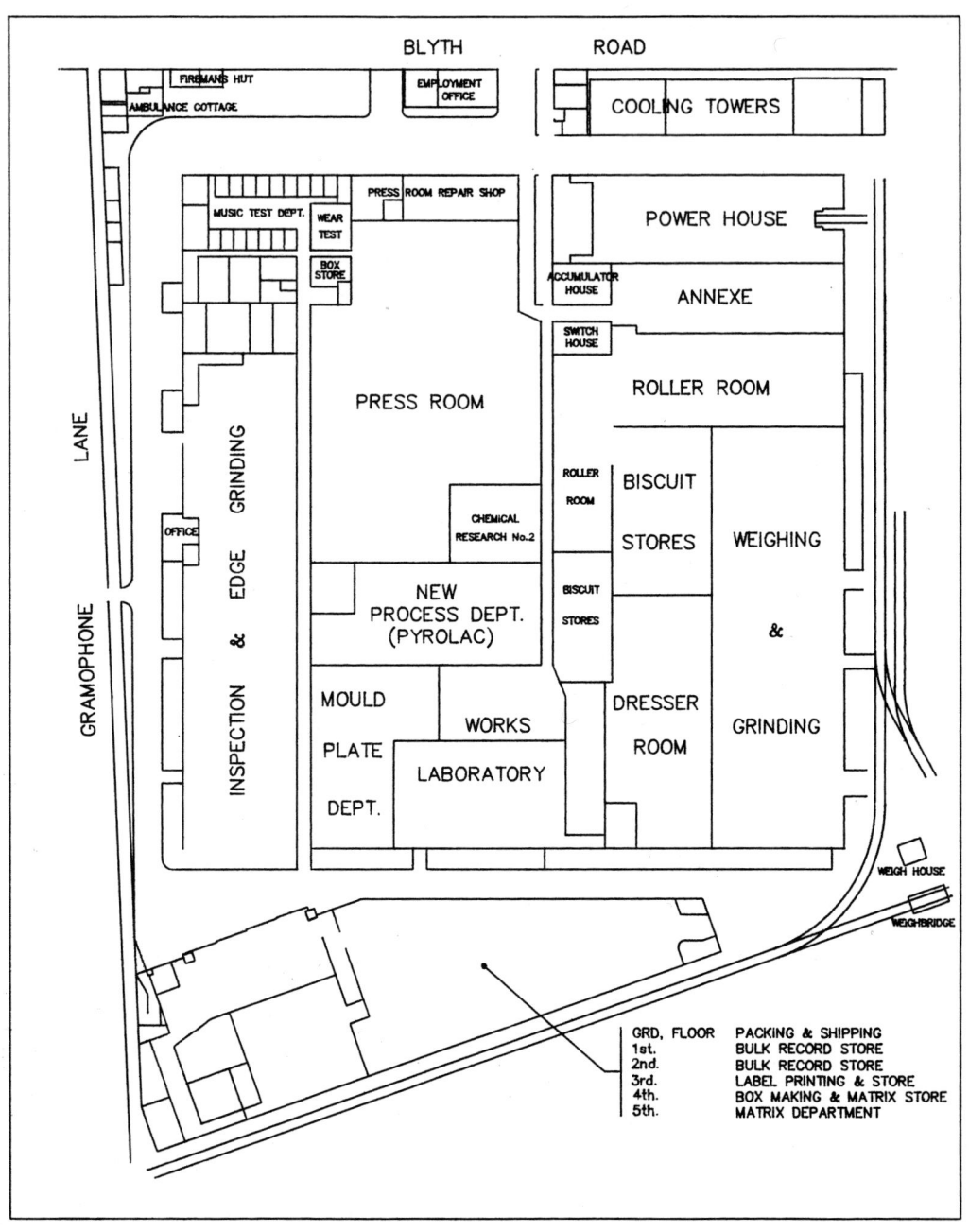

Blyth Road
Plan of the Record Factory.
(See page 216 for site plan of Uxbridge Road manufacturing facility.)

CHAPTER ONE

The Factory Sites

THE HAYES AREA AND ITS COMMUNICATION AND TRANSPORT LINKS

EMI's record factory has always been located in Hayes but where is Hayes, or for that matter the Blyth Road or Uxbridge Road sites, located?

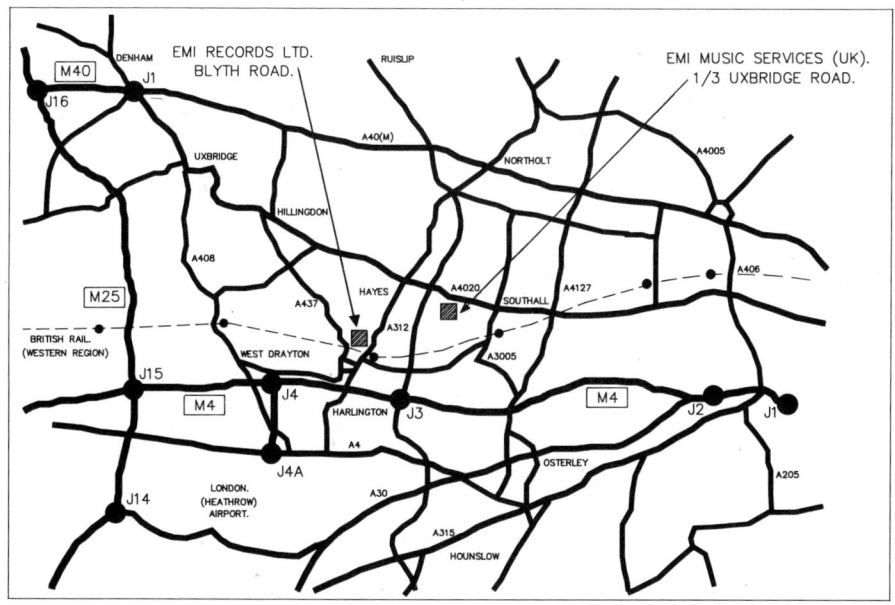

Diagram showing the location of the two factory sites and local communications.

The town of Hayes is about 13 miles to the west of London, in the county of Middlesex and is within the Borough of Hillingdon. The town has been associated with the name of The Gramophone Company, and later EMI, since 1907.

The borough includes within its boundaries part of the largest airport in England, London Heathrow. A panoramic view of this could be obtained from the fifth floor of The Gramophone Company's record stores building, in Blyth Road.

While the record factory was located in Blyth Road, the airport actually had little effect on working conditions, apart from the odd rumble carried on a southerly wind. This was not true, however, of the quality control listening equipment located on the fifth floor of the stores building. It was plagued by a rhythmic buzzing sound, corresponding to the rotation of the radar scanners.

Although the Uxbridge Road factory is further from the airport than Blyth Road, it is near the approach flight path of runway 15, which is used for landings when strong south-westerly winds are blowing. In this location the factory is not affected by the radar, but pilots calling the tower during final approach break through the Independent Radio News broadcast in the factory. These broadcasts were relayed on the licensed internal radio station. One compensation for this inconvenience is the magnificent views of aircraft on the approach, especially Concorde.

The Hayes area is well served by public transport, and close to the M4 and M40 motorways. Before the motorway network was planned several main roads converged on the area, bringing with them some traffic chaos.

Parking has always been a problem on the Blyth Road site. Indeed a letter from a Director of The Gramophone Company in the 1920s complains of the car park by the head office being filled by staff cars, some of which he intimates were illegally parked.

Blyth Road is close to the centre of Hayes town and the former Great Western station of Hayes and Harlington. This station is located on the main Paddington to Reading line and is normally served by local trains. In the past The Gramophone Company chartered "specials" to seaside locations for the annual staff outings. Numerous express trains thunder through on the fast lines.

The company never had a station of its own, but the main line station name boards were enlarged to hold the message:

HAYES & HARLINGTON
THE HOME OF HIS MASTER'S VOICE.

A complex of Company sidings connected to the main line close to Hayes Bridge. In the time that the record factory was located on this site EMI was to own at least two steam tank locomotives to handle internal shunting duties.

Aerial view of the Blyth Road site in the mid-1960s.

The main function of the internal railway system was to supply fuel to the power house boilers, whose capacity was such that they supplied services to all parts of the site, not least being the record manufacturing process.

A steam engine was also used to propel a trial delivery of ICI polymer [used in the record-making process] delivered by rail tanker. The tanker was shunted close to a storage silo, to allow the polymer powder to be transferred. Perhaps the fact that a "Presflo Cement" tanker was used might have some bearing on why this experiment was not repeated!

The latter of the two steam locomotives was supplemented by a diesel shunter, of indeterminate age, that previously served the Castrol oil terminal, located between Clayton Road and the Grand Union Canal. Occasionally this locomotive would give up the struggle of moving loaded oil tankers, and the steam engine would venture into the terminal with all its attendant dangers.

The works steam loco seen at the back of the record stores.

Blyth Road was used as a bus terminal point for several London Transport red bus routes. More than once the writer has witnessed a bus arrive and depart preceded by the conductor carrying a flaming torch. This was to guide the driver through one of the infamous London pea-souper fogs, caused by the burning of coal. Fogs like these had to be experienced to be believed. These fogs progressively became a thing of the past, following the introduction of the Clean Air Act.

THE BLYTH ROAD SITE

As was mentioned earlier, the 80-acre site consisted of many different factories, of which The Gramophone Company's was but one. Until 1955 some 16,000 people were employed here, but approximately 9,000 lost their jobs following Sir Joseph Lockwood's decision to halt production of domestic appliances, television sets and record players. The wooden cabinets for the latter were still, even then, being hand made. Let us explore the complex, and the buildings associated with it, by taking a walk through the site in the late 1950s. Later we will undertake a tour of the record factory.

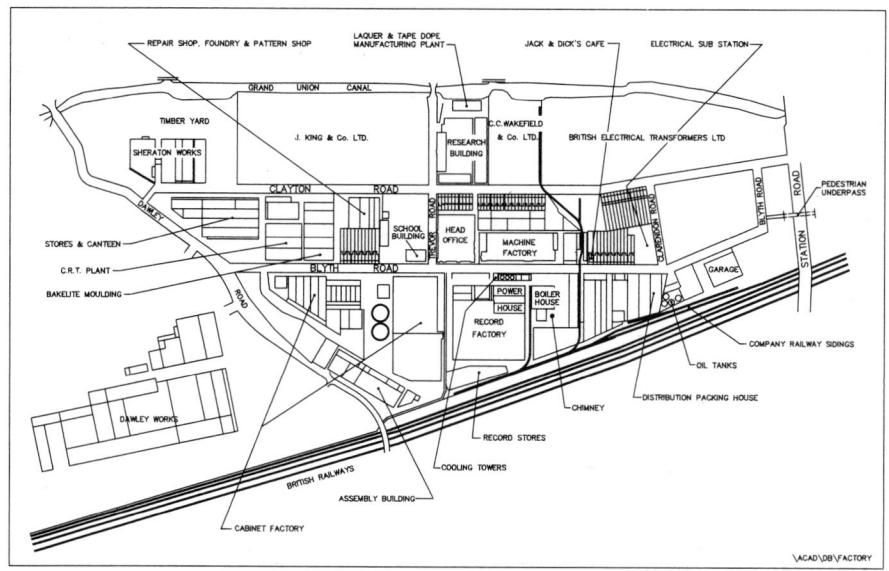

Plan of The Gramophone Company Limited, Hayes, Middlesex and adjacent properties. (circa 1955).

We commence our walk at the station exit and after traversing the pedestrian underpass our attention is immediately drawn to the huge brick chimney in the middle distance that dominates the view. On the left are the Company's garages with their imposing wooden doors. Here everything from the Managing Director's car to delivery vans were serviced.

Another task undertaken by the garage was the regular maintenance of the electric trucks that could be found hurrying hither and thither in all areas of the site.

While passing the garage we would have noticed a few bus stops, with their associated shelters. Had we been passing during the morning or evening rush hours, we would have appreciated just how much the EMI work force then made use of public transport. The opposite side of the road was occupied by terraced houses.

Reaching Clarendon Road we would have noted a further bus stop and shelter with a route 140, RT type, bus bound for Mill Hill Broadway, standing alongside awaiting its departure time. The double-decker bus is dwarfed by a tall building, which was the electrical sub-station for the whole site.

Looking to our left we would have seen the stainless steel oil fuel storage tanks that had replaced the earlier coal supplies for the power house. Beyond these lay the Company's sidings, giving access to the main line, and behind them a wooden fence. At the rear of this, a set of vintage suburban railway carriages were stabled most weekdays.

Continuing, we would have passed, on the left, an extensive brick building with high windows. This was the distribution packing operation where product was prepared for despatch to all destinations at home and abroad. To our right a row of terraced houses ended at the famous 'Jack and Dick's Cafe'. Famous, that is, to employees who arrived well before starting time to enjoy one of their giant breakfasts.

Our perambulations are now temporarily halted by a man with a red flag, halting traffic to allow the diesel shunter to cross the road with an oil tanker in tow.

On our way again, the five storey building on our right is the machine factory with its own huge water storage tank located on stilts above the building.

Water would be the main attraction on our left, as the huge wooden water cooling towers discharged enormous quantities of steam into the atmosphere. Depending on the prevailing weather conditions, this could be a spectacular display with visibility reduced to a few yards by the billowing steam. The sound of these coolers was quite deafening. Behind them stood the power house that generated power and steam for both production and heating for virtually the entire site. Generation at times exceeded site requirements and then a complex of switches would be thrown to feed power back into the national grid.

Next came the main entrance to the gramophone factory, which we shall visit later. Let us take this opportunity to look at the other buildings associated with the EMI complex.

Directly opposite the record factory was the head office building with a smartly uniformed Commissioner on duty at the main entrance. Before the recording studios were established in London at Abbey Road, most EMI recordings took place in the studios located within this building.

Behind the head office stood the research building famous for Schoenberg's development of the cathode ray tube. Like other major buildings on site, it had five floors above ground. It was physically connected to the accounts building, the top floor of which was given over to audio research under the leadership of Doctor Dutton. Within this department many experiments and discoveries were to lead EMI to the long play record and tape storage mediums we know today.

Behind the research building was another major landmark associated with television, a prototype transmission mast. Its height was such that red aircraft warning lights were placed half way up, as well as on top of the mast. It was dismantled in the early 1980s much to the relief of many locals, who claimed that it attracted lightning strikes! A section from the mast, complete with its warning light, is preserved in the grounds of the new research building, Central Research Laboratories.

At the rear of the accounts building stood an anechoic chamber used for experimenting with loudspeaker design, and close by stood the lacquer and tape dope manufacturing plant. Both buildings were located beside the Grand Union Canal.

Final demolition of the old Personnel Building occurred around 1962. The 14 HMV heaters, used to heat this building, became obsolete!

Beyond head office on our right was a building that looked distinctly like a school, and that, indeed, was its original role. Now looking rather dilapidated, it had previously been occupied by the personnel department. It was partly demolished some years later by driving an experimental tank through it. Space made available by this act was immediately utilised as additional car parking!

At the rear of the school, a complex of single-storey buildings stretched through to Dawley Road. They included, in the foreground, the repair shop, the pattern shop and the foundry. The cathode ray tube and bakelite buildings occupied the centre area. In the distance the large works canteen was next to the general stores. Beyond this, the Sheraton works complex was bordered by Dawley Road, the Grand Union Canal and the premises of J. King and Co. Ltd.

Looking the full length of Blyth Road one would have seen the facade of Dawley Works, originally owned by the Rudge Whitworth Bicycle Factory, which was the only part of the site not connected to the internal railway system. Dawley has a direct connection with our narrative as it was in laboratories at the rear of this complex that experiments and early production runs of EMI magnetic tape were conducted. A few years later a factory dedicated to magnetic tape manufacture was built here, in collaboration with Philips.

To our left now was the personnel department, with the five-storey electronics building in the middle distance beyond. Within this complex the professional tape machines and fader desks used in the factory and studios were designed and built. In later years this operation was also responsible for designing and refining a revolutionary disc testing apparatus known as the Disc Defect Detector. These were used in many record factories, worldwide.

There was little communication between the individual factories. This did not hinder the camaraderie between workers in the various buildings, as on a journey between them, or to the canteen, one was bound to meet staff from the other operations. Being close to the town centre also meant that a shopping expedition could be undertaken during the lunch time.

Retracing our steps we would have entered the front yard of The Gramophone Company's record factory where an ambulance and fire engine would be standing outside their respective garages. The ambulance is associated with the surgery, which was located near the factory entrance. Immediately inside this entrance stood the police post and this is where we present our credentials and await our guide who will take us through the complexities of record production.

A TOUR OF THE BLYTH ROAD RECORD FACTORY

[Before reading this section, please remember that we are being guided round the factory in the late 1950s.]

The initial impression one has while waiting for our guide is the silence, a false one as we are to find out. When he arrives, introductions are exchanged and we are welcomed to the EMI record factory. In an office we are offered the normal pleasantries, followed by a resume of what product the factory is producing at this time.

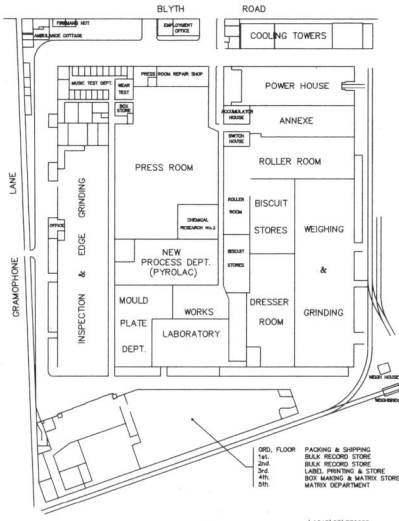

Plan of The Record Factory (circa 1950).

These are listed below:

> 78 rpm shellac discs, in both 10 and 12 inch versions.
>
> 45 rpm vinyl records, in both standard play and extended play versions.
>
> 33 1/3 rpm vinyl long play Monophonic records in both 10 and 12 inch versions.
>
> 33 1/3 rpm vinyl long play Stereophonic records in both 10 and 12 inch versions.
>
> Reel-to-reel pre-recorded tapes.

Our first question is about record speeds and sizes. Simple? No, as we shall learn!

An early photograph of visitors at Blyth Road, watching 78 rpm records being pressed.

We are told that the common speeds of disc rotation are 33 1/3 rpm, 45 rpm and 78 rpm. However in the early days of recording, when considering the effect of the crude driving mechanisms available, the speed of a colloquially termed 78 rpm could vary from about 55 rpm to 120 rpm. [Now that old masters are being used to re-release historic recordings, the use of a variable speed turntable is imperative. However, by the time of the First World War the speed of the 78 rpm was constant.]

Until recently, records rotating at 22 rpm were produced in the factory with a braille label on one side. They had talking books for the blind recorded on them. Although not utilising microgroove technology, the speed of 33 1/3 was used until recently for certain BBC and film sound track recordings.

With the recent launch of 33 1/3 rpm and 45 rpm microgroove records, their speeds are now regarded as the standards. Experiments with slower speeds, e.g., 16 2/3 rpm, are being conducted. They are not thought to be commercially viable, however, and would only be used for speech recordings, as their frequency response would be too poor for music reproduction.

That was not as simple as I thought it would be, what can you tell us about record sizes?

Record diameters, we are told, have varied greatly during the history of recorded sound. The largest discs produced by The Gramophone Company were 16" diameter, while the smallest were only 2" diameter, and were made especially for the Queen's Dolls House. They are now on public display in the State Apartments at Windsor Castle. Incidentally, these discs play at 78 rpm, and they are recordings of a brass band playing various anthems, each of 7 seconds duration. 5" and 6" discs had also been manufactured in earlier times. Nowadays we recognise 7", 10" and 12" diameter records as standard.

It is felt that we have asked enough questions at this stage so we begin the tour. Before returning to the ground floor we are introduced to the Factory Manager, Mr. Harry Christmas, and Mr. Wally Rand, Engineering Manager.

In the main factory corridor the silence of the entrance foyer is replaced by the rumble of machinery and the occasional snatch of music. It is explained that the music is coming from the newly built individual quality control listening booths. These are on our right, while the sound of machinery comes from the pressroom and toolroom, which are on our left. Leaving the main corridor we enter an outside alley-way, which is overshadowed by the five storeys of the record stores and cabinet factory buildings.

Our tour is to commence on the fifth floor of the record stores building, in the matrix department. To get there we travel in a large goods lift with an attendant who works the heavy doors and calls the floor numbers. To give some idea of the size and carrying capacity of this lift we share our journey with one of the electric trucks seen during our walk round the Blyth Road site! On arrival, we go around the outside corridor to meet Jack Wheeler, the Matrix Department Supervisor.

After being introduced, we enquire if this really is the start of our journey? Yes, we are told, at least as far as the factory is concerned. Since the Abbey Road studios were opened, all commercial wax and lacquer cuts are produced there.

Two cutting machines, or lathes, do exist at Hayes; one is located in the audio research department, while another, older machine is located within the works audio laboratory. It is colloquially known as a "bacon slicer" for reasons that, we are told, will become obvious when we see it. Experimental cuts made on both machines are regularly processed. The research department machine has recently been fitted with a stereo cutting head.

A wax is about to be cut on the custom built lathe, photographed, probably, in the head office in Blyth Road.

What then, is a wax or lacquer? Ever since Edison discovered that sound has a distinctive wave form, by tracing its wave pattern onto a revolving smoked glass plate, experiments have been conducted into ways of permanently preserving it. The method employed throughout most of the shellac era was to cut the wave form into wax.

With the coming of the microgroove record the lacquer was adopted. This is a cellulose based solution coated on to an aluminium blank, which is dried in an oven to give a very high viscosity surface, into which the wave form can be cut. Some of the waxes and lacquers processed in the department are manufactured at Hayes. Lacquers are coated on to various size blanks ranging in diameter from 16" down to 7". Waxes were normally flow coated on to 13" glass plates.

The first, and most important, action in either process is to engrave or stamp the lacquer or wax with its matrix number. This is its individual reference, which it will carry for the rest of its life. When the lacquer was cut at the studios the catalogue number, side and cut number is scratched on to it's surface, outside the playing area. Each set of lacquers delivered from the studios is accompanied by a packing slip that also gives these numbers. A matrix progress card accompanies every job throughout the process, each operator signing for the operation they perform.

The silvered disc is prepared for growing in a special holder.

We enquire why the matrix numbers differ from the number that is given to the record retailer? Historically this has always been the case, but there are proposals to make them the same.

At this point we are introduced to Len Morton, who will guide us through the process. At the time of our visit both 78s and LP's are being cut into the lacquer medium noted above. The lacquer is non conductive and, as it requires to be electroplated so as to grow a master, it has to be made conductive. This is achieved by chemically reducing silver on its surface. After being immersed in a wetting solution the lacquer is placed in a spray unit where the silver is deposited.

We are shown the plating cells, and at first glance they look like something that the three witches in Macbeth might have used! They are steel vats that are lined with rubber, and contain a foaming green liquid which, we are told, is the plating solution, nickel sulphamate. At the base of each cell are baskets of pure nickel. This nickel forms the anode in the plating process. The cathode is the part that is to be plated, which revolves partially immersed in the plating solution. A DC electric current is passed from the anode to the cathode, which deposits pure nickel on to the part being plated.

We observed a silvered lacquer being attached to a cathode spindle, and lowered into the solution. It was explained that this process takes four hours, and produces the first part, the master. In another bath a lacquer has completed this process, and after being thoroughly washed in water, we observe the two components being parted. The outer edge is physically cut on an antiquated looking wheel trimming machine. The two parts are now separated, to reveal the nickel master.

Lacquers are stored in case of quality queries, but are unlikely to be used to make a second master. We have now created an exact negative of the original, which is handled with great care as the groove stands proud of the surface. The matrix number inscribed on it is a mirror image, and not easy to read. To avoid confusion the matrix number is written on the reverse, and covered with transparent waterproof tape.

The master is then placed in another unit, and sprayed with a colloidal filming solution. This will allow the next metal part, which is called a positive, to be separated from it after growing. The master is plated in a cell identical with the lacquer plating unit, the only difference being that higher amperage is used from the start of the process, rather than being increased in steps. This is possible because the component being plated is a solid metal part, and will not, therefore, suffer heat distortion as the lacquers would.

When the growing process is completed, after approximately two hours, the combined metal parts are washed in water and dried before the separation process can proceed. To the uninitiated, this operation looks extremely

Another metal part is put into the plating bath for growing.

dangerous, but is safely handled by a skilled operator. A separating knife is placed against the edge of the combined parts and pressure applied. This breaks the seal, and the knife can now be drawn around the edge. The two are separated, each being placed face down on plastic trays.

The master, after further cleaning, is packed and placed in the store, where it remains until further positives are required. The positive is trimmed to a standard diameter, depending on the size of the finished record, and passed to the matrix quality control department, where it is visually inspected and played.

Before visiting the quality control, laboratory and pilot plant areas we are shown the "dry area", which is not a prohibition zone, but literally an area where plating solutions are not used. Here we see the positive being optically centred before being aurally checked.

In matrix quality control we meet Mr. Dewar who invites us to listen to what, until now, has just been identified by a matrix number. Betty, a Music Test Operative, plays the positive throughout, listening both to the programme content and for any faults that may have occurred in the process. We are fortunate that the number we have been following is a stereo matrix. This gives us the opportunity to assess the difference from monophonic or single channel recording, and we are suitably impressed. The cyclic buzzing sound, caused by Heathrow's radar, can be clearly heard through the loudspeakers, but the operator ignores it. On this occasion the result is satisfactory, but what happens if it's not, we ask?

Mr. Dewar takes us to a room known as the repair booth. Microscopes are arranged round the benches, with white-coated technicians looking into the eye pieces. On the turntable under the objective lenses are metal parts. Between the operator's fingers is a rod with a thin wire on the end. It is explained that this section's job is to assess the severity of faults and, if practical, repair them, or have new lacquers ordered. The job is obviously very skilled. The repair consists of physically removing or wearing down protrusions within the groove.

Before returning to the plating process we are shown the matrix laboratory, which is under the control of Ray Burnett. This is where the chemical composition of the plating solution is monitored. Checks on the bath's acidity, known as its Ph., and the density of the plating solution, are made twice daily. The laboratory is also responsible for preparing the various colloidal filming solutions used in the process.

Our eye is caught by a large glass container of boiling water, with a water cooled condenser above it. This, we are told, is producing distilled water used in the preparation of the various solutions. As the

This suitably attired operator, with protective apron and gloves, removes the grown metalwork from the bath.

quantity of water used, as a rinse in the spray units and to control the bath densities, far exceeds the output of distilled water from the laboratory, a demineralised water plant has been purchased. This plant is similar in action to that of a water softener.

Concentrated plating solution is bought in bulk from a supplier, and diluted to a suitable density using demineralised water. Next door to the laboratory is a pilot plant. This consists of three individual plating units, a chrome plating bath and some other vats, one of which, we are told by Bill, the technician in charge, holds some twenty gallons of cyanide solution. The main reason for the plant's existence is to conduct experiments with plating techniques before they are used in production.

The positive we have been following has now been returned to the production floor. It is again placed in a spray unit, to be treated with a colloidal filming solution, so that the positive and its derivative, the stamper, can be separated. This takes about one hour in the plating bath.

The separation procedure is the same as before, the stamper being treated with care, as it is again a part with ridges rather than grooves. Positives are either stored or returned for further stampers to be grown. Our stamper is sent to the dry area to be back sanded to remove any roughness. Next it is centred using an optical microscope

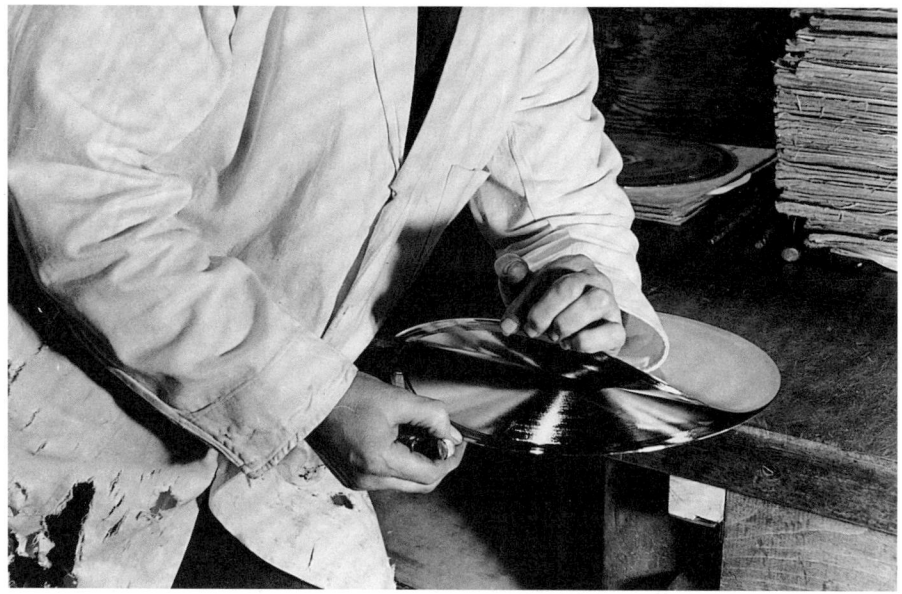

The positive is carefully separated from the master. The effects of the acidic solution on the operator's coat is apparent!

combined with a punch mechanism. Now it can be pre-formed in a power press, which gives it the profile of a record press mould block. Finally it is visually inspected and given a stamper letter.

It was explained that The Gramophone Company used a letter coding based on the Company's name, i.e.,

G. R. A. M. O. P. H. L. T. D.
1. 2. 3. 4. 5. 6. 7. 8. 9. 0.

A stamper marked GO is, therefore, the 15th stamper produced from the same positive.

We watch our stamper being packed into a strong cardboard envelope, on which the matrix and catalogue numbers are written. The original document is replaced by a Stamper Identification Card, which will accompany it for the rest of its life. It is now despatched to the stamper stores to await a pressing order.

While we were being shown round the dry area we had noticed three record presses, one each for 7", 10" and 12" discs. These are not refugees from the main pressroom we passed by in the corridor, but are utilised to manufacture test pressings. They have enough work between them to keep a full time pressman employed, and we are introduced, for him to explain his part in the operation. Our first question is why have three separate presses? He explains that each press is fitted with mould blocks to suit the size of disc being produced. After manufacture the test pressings are first played by matrix quality control. When they have been approved they are sent to the artists and managers, who listen to check that the sound they required has been faithfully reproduced. With the coming of stereo there are other things to check, including positioning of the artists, relative to the "left" and "right" loudspeakers.

What about test 78s we ask, and are shown at the far end of the floor two presses awaiting disposal to an overseas manufacturing plant. These were, until a few months ago, the 10" and 12" 78 rpm presses, which, following cut backs in releases of these products, are now no longer required. When test pressings for 78s are now wanted, they are made in the main pressroom.

All test pressings are known as "G" copies, as they are usually made from the first stamper produced from the first approved positive [see table on previous page]. They are generally single-sided, with a non-commercial stamper on the B side, and are made with a "plain" or "white label", the alternative name for such discs. A copy of each is stored in the matrix department. When all parties are satisfied that the disc is commercial, its metal parts can be "released" - that is, the stampers can now be sent to the pressroom when an order for the particular coupling number is raised. The coming of stereophonic

The label stores.

recordings has meant that there is a lot more work for this department, as they now have to originate both mono and stereo versions of each new release.

We then return to Len Morton, who takes us to be introduced to Jim Perryman, the foreman of the label printing department and metal work store. He will show us round his department. Also associated with the stores is the cataloguing department, and we are told that we will examine their role in the process later.

First we visit the metalwork store, where the masters, positives and stampers are kept in alphabetical and numerical sequence. The masters are stored in a fire proof vault, although the other two components are placed on open racking. To our eyes it seems impossible to locate a required piece of metal work in here, but to the trained technicians who work in this store it is simple. It is controlled by extremely accurate, manually maintained records.

Each lacquer or wax delivered to the matrix has its origination paperwork passed down to the cataloguing department. They had already allotted it a matrix number [different to the record catalogue number], which had been sent to the recording managers office.

Following receipt of the originating paper work a hand-written card, which we saw earlier, is released. If, however, the wax or lacquer is a new release, a catalogue card is also created for it. In the case of a recut for an already released item, the original card is updated to show the date on which the original metal work was destroyed, and the new cut received.

The metalwork store receives an addition to stock.

As has been noted elsewhere, both Hayes sites were known as the "mother" factory, and often supplied masters or positives to overseas factories. The department also updated the catalogue card with the catalogue number of the disc in each country it was to be released in. This arose because, while various countries might issue the same recording, they often did so under different catalogue numbers.

The packing of metal parts for overseas was undertaken in the distribution packing department, which was noted earlier on our walk down Blyth Road. To the casual observer, such an operation seems simple - just protect the playing surface, pack it and despatch it. This was, of course, fundamentally the case, but remember we are looking at the process in the late 1950s when today's sophisticated and environmentally controlled conditions were not available.

EMI had factories in many remote parts of the world, and many metal components had to be transported in environments that would not be encountered in Britain. So apart from getting scratched, the surface could grow a kind of mould or, in very extreme conditions, start to flake. The department charged with the packing process had much experience in sending product all over the world, and was able to package the masters in a manner suitable for each journey!

Now we move on to the label printing department, located on the third floor. This is where labels are printed for all EMI releases. Label information is printed on to paper with the backgrounds, such as the famous dog and trumpet logo, already on them. As with the matrix department, the work load has been increased by the simultaneous release of both mono and stereo couplings for the same recording, requiring two different labels for each release.

We are shown the compositing room, where label copy, supplied by the marketing company, is transformed into the record label information. The label is proof-read, using the original copy as a comparison. Labels are stored by catalogue number, rather than matrix number. At this point we hear a strident tone coming from a loud speaker suspended over our heads, and the more distant sound of a steam whistle; it's the noon time-signal.

Returning to the fifth floor, we meet Len before walking round the corridor to the lift. On the way we hear a strange beating and whirring sound. All three of us are intrigued and look out of the windows to see a helicopter hovering, before landing gently in the grounds of the Fairey Aviation building directly opposite us. We are told this is a fairly rare sight, even for those who work in the building every day. We wonder if any more surprises await us, and soon find the answer is yes, as a notice on the lift door informs us it is out of action for maintenance. This, we are told, is not such a rarity as the helicopter! The walk down five flights of stairs will give us an appetite. On the way down, our guide explains that the first and second floors are occupied as record stores. The ground floor is used for packing and shipping.

Reaching the ground floor we walk in the open air towards the head office, and walk up Blyth Road toward Dawley. Crossing the road we enter the roadway leading to the works canteen.

The canteen where workers enjoyed a meal break. Note the sandwiches brought from home and the knitting.

After an appetizing lunch we leave the canteen and walk back to the record factory, to visit the works laboratory. This, we are told, is not only a place where experiments are carried out, but also contains many small sections, replicating virtually every department in the factory. We are rather glad that the laboratory is on the ground floor, as the lift is still out of use.

Before our guide leaves us, we are introduced to Mr. Foxcroft, head of the works laboratory. Following a cordial welcome we are introduced to Mr. Jim Hughes, who will escort us round the laboratory complex. It is explained that, although the laboratory is located in the record factory, its mandate is to supply information to all branches of the company, and, where practical, give advice to overseas factories.

First we meet Mr. Parker who, with his team, is responsible for the study of metallurgy. As was noted earlier, the site includes a foundry, and the testing of metals produced there makes up a major part of their work load. They also examine metal parts produced in the matrix department. The chemical section, under Mr. Hughes, undertakes a varied menu of tasks, ranging from routine analysis through to specialised projects for both the Hayes site and overseas factories.

At this point our noses are assailed by a distinctive acidic smell; Ian, an assistant, is checking how long an experimental material takes to decompose. This is called a congo red test, and the time the material lasts is ridiculously short. Pity, Mr. Hughes observes, for the audio quality of it was satisfactory. They have, in fact, been testing a material for its potential to be used to make vinyl records.

We meet Mr. Perren and his assistant Audrey, who have charge of a sophisticated electron microscope. This area of the laboratory also has three staff to wash equipment, and do general cleaning duties. The clerical staff consist of Miss Perks, secretary to Mr. Foxcroft, plus three full-time secretaries. All letters and reports had to be typed, with copies being transferred by carbon [no photo-copiers in those days!].

The last section in this area is not actually part of the laboratory, but comes under control of the audio section. As it requires a dark room, however, it is located here. We are introduced to Chris Pooley, who has mounted a worn record stylus on a shadow graph microscope for us to view. He explains that the image we can see on the screen is a magnified view of the stylus showing flattening caused by the friction generated when it is used. The stylus in question has been in use playing positives in the matrix quality control for just 24 hours. Having been rejected it will be sent away to have a new diamond fitted. Normally each magnified image is photographed, as analysis of styli wear is being conducted following the introduction of stereophonic records.

Mr. Hughes explains that he now has another engagement, and asks Chris to continue with the tour in the other half of the laboratory, in the record stores building, opposite. The first area we enter is the experimental roller room, where machines mix and roll out record material. We meet Roy Matthews [later to become Factory Manager] who is in charge of this experimental area. Tests and experiments undertaken here include routine checking of the compound being produced by the main compounding area, known colloquially as the black room.

Routine checks on deliveries of constituents in the standard compound are also carried out. This is performed by substituting a sample from the new delivery in an already approved batch, and then testing it against the standard.

Experiments are, additionally, conducted with new compounds, many sent in by companies hoping that some by-product from their manufacturing process might find a use in the record industry. They rarely prove satisfactory either from a chemical or audio point of view. Again overseas factories were mentioned, as they regularly send routine and experimental materials for analysis.

We shall return to this area later when the technicians return from lunch; at this moment the one o'clock time signal sounds.

Next door is the experimental paints laboratory under the control of Percy Evans. They don't have a great deal to do with record production, but do analyse new carbon black pigments. He says he expects that in the future people will get fed up with black records, and then they will be busy testing additives to create other colours. [As time has shown black is still the predominant colour, with only the occasional coloured record being produced.]

Can you tell us why records are black? The simplest answer is that they always have been, and from a practical viewpoint it makes the spotting of faults much easier. To prove the point we are shown a disc made from a formulation with no carbon black. The first thing we note is that the groove can be seen from the other side, through the disc. Additionally, when placed on a turntable it is difficult to find where each band starts.

Moving on, we enter the experimental tape coating area, where our guide points out some large stone jars that are being rotated on a machine to mix the contents, thoroughly, before use. These contain round stone balls and experimental liquid tape compound. The area is under the control of Ken Gray and his team, who are responsible for testing routine and experimental tape compounds [the coating that goes on P.V.C. film to form the basis of recording tape]. Inside the room we see a small coating machine and note that all the power and light fittings are of a fire proof design. The reason for these precautions is that liquids with very low flash points are utilised in the process!

The room is also used for the manufacture of small quantities of experimental and routine lacquer formulations. After mixing these formulations, the resultant liquid is dark purple in colour, having the same consistency as treacle. It is subsequently stored in glass bottles. To be tested, it has to be hand-coated on to aluminium blanks. Our guide tells us that this is a difficult process, as air tends to get trapped during the coating process, which, as the liquid dries, forms blisters in the surface.

When dry the lacquers are cut on the machine that is known as the bacon slicer, which we are told we will see shortly. Our attention is diverted by a deep rumbling sound that seems to be coming from outside. No-one else seems to notice, but all is revealed when the works steam engine passes by a few feet from where we are standing! We thank Ken and his assistants for their time.

Next we visit the audio laboratory, and here we meet Terry and his team who will show and, we hope, explain, some of the mysteries of the gramophone record. As we enter the department our eye is immediately drawn to a record player with three playing arms! As we watch the most amazing thing happens, for, after covering the first three bands, all the arms lift, one after the other, and return to the area of the "run in". After the third arm has settled the record continues to be played. We have seen many items of interest today, but surely this is the most unusual!

The machine we have been watching is used to assess the wear properties of both routine and experimental long playing record materials. This is achieved by subjecting the disc to a minimum of 60 plays using a needle weight of 6 grammes. A disc manufactured from standard production material is wear tested as a "control" for comparison with the material being evaluated. Chris forestalls our next question by telling us that 78s are also wear tested, and we shall see this later, when we visit the finishing department.

Now to see the famous bacon slicer. We are introduced to Mike who looks after the laboratory cutting equipment. The first question has to be why call it a bacon slicer? To explain, Mike operates the machine for us. A lacquer is placed on the turntable, and secured in place. He then turns on a vacuum cleaner, which is connected by a series of tubes to a collection jar, and from there to a tube that ends just behind the cutter. This apparatus draws off the swarf created as the lacquer is cut. It is a highly flammable material, and the collection jar is regularly emptied into a fire proof bucket, the contents of which are disposed of by the fire department!

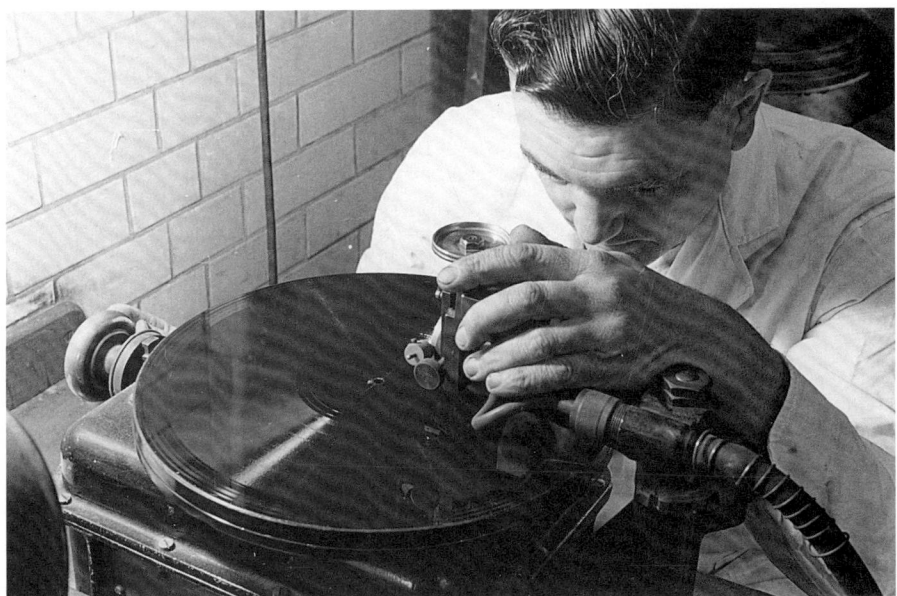

Cutting the run-out groove.

Next a motor is started, which is connected by several pulleys and belts to the turntable. This sets the revolutions per minute of the cut, and all its subsequent copies. On older machines this motor also powered a worm gear that, when engaged, drove the head sideways across the lacquer surface at a set speed. Therefore the cut has a known number of revolutions in each inch of the lacquer; this is known as the turns per inch [TPI]. A problem that arose from the older cutters, however, was that if there was loud modulation on the recording, the groove would become wider, and tend to run into the groove next to it.

Mike explains that since the introduction of magnetic tape it has been possible to preview the volume of the music by means of a second replay head, placed first in the head chain. The signal from this is fed via a special amplifier to a motor, which has replaced the worm gear. So, if the head senses loud modulation, it speeds the motor up giving more space between each groove. Conversely, lower levels will slow it down, bringing the grooves closer together. This is known as variable groove spacing.

The machine located in the research building has this facility, as well as a push button to activate faster movements of the head across the surface, for the run in, cross overs and the run out. It also automatically lifts the cutter after just over one revolution of the lock groove. Prior to the availability of this development, cutting 78s was a very tricky business, and required a high level of skill from the cutting engineer.

Before the cutter on the laboratory machine is lowered on to the lacquer, we are shown a small coil of wire, which is wound round it. When energised it heats the tip, thus helping the cutting process.

The cutter is lowered onto the lacquer. Once the cut has been started, we are shown that to speed up the rate at which the head is driven across the revolving lacquer surface, one has to turn a wheel. This reminds one strongly of bacon slicing machines, which were to be found in grocers at this time, hence the machine's nickname! A steady hand is essential to give an even look to the scroll.

We are invited to try, and find that it is not easy to get an even look to the grooves. At the end of the side the lock groove has to be judged very carefully, as the lacquers thickness only allows for a maximum of a double cut. Cut too shallow and the groove does not lock, too much and the cutter goes through to the aluminium base!

We are invited to look at a stereo record groove under the microscope. As the record is turned we see the variable spacing referred to above, and it is also possible to see indentations on the sides of the groove wall corresponding to the stereo modulation. The groove is deepest in the heavily modulated areas, becoming shallower in quiet areas. Terry asks if we think that stereo is a new invention? We reply yes, and learn that a patent for a stereo system was taken out before the Second World War, by an EMI inventor, Mr. Blumline. The only difference between his method, and the one we are using today, is that the depth of groove was used for one channel, and the other was recorded on the groove walls. This was known as the hill and dale method.

We continue to look at the groove and note a long indentation in the base, or apex, of it. It is explained that this is a recently discovered phenomena, which, although it must have happened in the mono era, has now become a major problem with stereo records, as it can be heard on them. It is caused by the top and bottom stampers momentarily touching in the press, before the record material can act as a barrier, and is called an apex damage.

Have any other problems been highlighted by the stereo record we ask? We are told about a recently discovered fault in the stereo recording itself, known as phasing. This is shown to us. First an in-phase stereo record is played giving a clear sound image and clear placing of the instruments. Next an out-of-phase recording is played, and this gives indistinct placing and sound. A warning is given that the same effect can be obtained by reversing the polarity of one of the two speakers.

The ringing of a telephone bell interrupts any further conversation. Chris informs us that the plastics section have everything ready for us to see, a practical demonstration of the material mixing and rolling techniques. We return to it, and are introduced to Don, who carries out this task. The first, and by far the largest constituent [approximately 98.5%], of the mixture is the basic co-polymer. This feels slightly granular when rubbed between the fingers, and has a smell faintly reminiscent of ripe strawberries. The remaining 1.5% is made up of stabilisers, lubricants and carbon black. In this instance one standard lubricant was left out of the mix and substituted with a brown-coloured paste that had been supplied by some hopeful company.

All the ingredients are placed into a mixer that stands about five feet high. The lid is closed and the mixer switched on. The heat created by the mixing process is dissipated by a water filled jacket. After mixing, the dark grey coloured powder is poured into a paper sack. This is then taken to the first set of rollers located in the experimental roller room. Here we are introduced to Les, who, having removed a sample of the powder for analysis, went on to show us how the process worked.

The temperature of the steam-heated rollers was checked with a pyrometer, before the mixed powder was shovelled on to the revolving rollers, any dust being drawn up and away from the area by a large extractor hood. As the powder heated on the rollers it became a thick

The real thing in the weighing and grinding department.

viscous mass, which was cut at intervals, and thrown back on to the revolving rollers to complete the process. Finally a scraper blade was pressed against one roller to remove the material.

This was then transferred to another set of steam-heated rollers, which rolled it out to be approximately six feet long by one foot wide, with a thickness of about 3/16". This sheet was deeply scored, and when the material had cooled and hardened, these marks allowed it to be easily broken. The resultant material blocks were known as biscuits, and one was required for each 12" pressing.

When the process was completed, and the machinery switched off, one became aware of the silence, for the rolling process was not a quiet one! So we move from one noisy environment into another, in the experimental plastics laboratory. Mr. Matthews explains that the record press located here has its own hydraulic system supplying the pressure to form the record, and this is where the noise originates. Don has just finished a job and has turned off the hydraulic power, and removed the stampers. We now have the opportunity to view the complete process.

First, labels, stamped with the formulation number and a pressing sequence number, are placed in a steam-heated tray beside the press. This removes any moisture absorbed since they were previously dried and sealed. Then the mould blocks inside the press are coated with a waxy substance, which is carefully rubbed over the surface, and then polished. A test stamper, having the same programme as was mentioned in the wear testing of records, is cleaned and placed on the top block. The bottom stamper is then positioned in the press; it is one with a continuous tone.

The biscuits of material are placed on a steam-heated platen, where they become viscous again, and can be rolled up, ready to be placed in the press. Now that the labels are dried, they are placed in cups either side of the press, and Don cycles the press a couple of times, without the hydraulic pressure being applied, to "warm it up". The noisy hydraulic pump is now switched on, and three records with no labels are produced. These are made to check that no blemishes are apparent on the stamper. We ask what can cause such problems, and are told that micro amounts of dirt trapped under the stamper can cause bumps, which appear as small pits [known as sinks] on the finished disc.

No faults are found, however, and we note that the material is removed in a clockwise direction from the table with a fresh biscuit being put in its place. By the time five pressings have been produced the sixth biscuit has softened sufficiently to be rolled up. For this demonstration material that has been mixed for normal production, and made in the black room, is being used.

We are asked if we would like to produce a record! The first thing we have to do is place a label on the top and bottom stampers, but with the stampers being heated by steam, one feels that a set of asbestos fingers would be a distinct advantage! This being accomplished we now have to handle the viscous and extremely hot material. Don takes pity on us, however, and rolls it up for us, remarking that if we think this is hot, we should try handling the 78 rpm material! We now have to transfer the roll of material to the middle of the lower label. The way to do it is quickly, so that the senses don't fully comprehend the pain the heat is causing you!

This being accomplished buttons on each side of the press are depressed and the stampers come together gently. Although we cannot see it, the material is now spread over the stampers by the hydraulic pressure. After the cycle is completed the press opens to reveal the record, surrounded by excess material, hanging from the top stamper. We remove it by holding this excess material, and realise that it is hot but not uncomfortably so. The reason for this is that water has been passed through the mould block to accelerate the cooling of both the stamper and record.

Next we place it on a machine that cuts away the excess material [known as flash], utilising a turn table and heated cutting knife. Following this operation we remove the completed pressing from the trimmer, and are told that, as it has no commercial value, we could keep it, and to pick it up from security as we leave.

Finally we ask Don what happens to vinyl records that are no longer required. He explains that a commonly held theory is that they are melted down, but this is not so. The centres are removed, as no process has yet been invented to remove the paper labels satisfactorily, and they are then broken up and reprocessed into biscuits for 7" records.

Our original guide has now rejoined us for the final part of the tour. Before departing we thank our hosts in the plastics laboratory, and Chris, who reminds us to watch out for him later.

The next area to visit will be the pressroom followed by the examination department. It is the Examination Department Assistant Foreman who has kindly invited us to three o'clock tea. As we walk into the factory proper the "Calling all Workers" signature tune can be clearly heard. We are introduced to Mr. Yeoman, and spend a most enjoyable 15 minutes talking, not about gramophone records, but about railways in general and steam engines in particular!

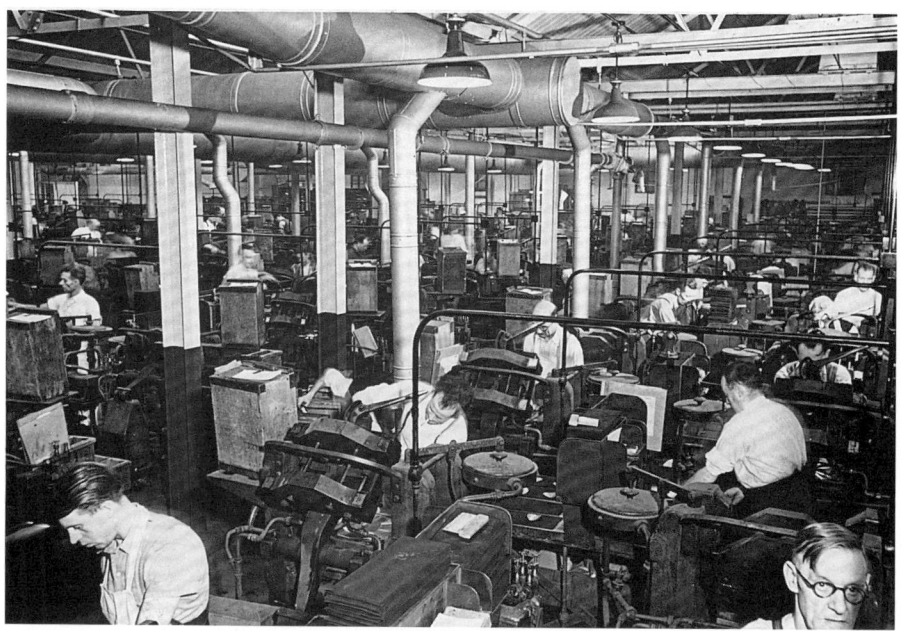

A hive of pressroom activity in the heyday of the 78's.

With the tea break over, we pay a visit to the pressroom and realise that our efforts in producing one record pale into insignificance against the scale required for commercial production. Our guide points out that, like everything else in life, training, practice and bonus schemes, in the form of piecework, are great motivators.

The pressroom itself is a vast area that is noisy, hot and very humid, with an interesting variety of both sounds and smells. In all, we are told, there are some 180 presses, and when they are all running it is even hotter and noisier than it is now! The presses are arranged in six double rows, three each of 7" and 12". Each press has numerous pipes projecting from the underfloor main services.

The presses are arranged in pairs, with one operator being assigned to each pair. Each pair of presses has, nearby, a large metal steam-heated table, above which there are infra-red heaters. It is very hot beside these machines! They are used to pre-heat the biscuits of vinyl material, which are laid on the tables, before pressing.

We are told that the presses are semi-automatic. The operator loads the press with labels and the hot biscuit, rolled cigar-shape, and closes the press by means of air-operated valves. The press then takes

The operator inspects the pressing and "Long Tom" lubricates the press. A picture capturing the atmosphere of the time.

over, and cycles through periods of steam heating and water cooling, while under a pressure of one tonne per square inch. At the end of its cycle the press opens, and the operator removes the moulded disc and starts again. Between cycles the operator also has to trim the records, using a turntable and hot knife arrangement on 12", and an air-operated blanking tool on 7". He also has to inspect the record - piecework was only paid for good records! The records are then stacked on a spindle, or "stand". The operators all appear extremely busy.

Our attention is drawn to a separate huge press, unlike any of the others. It is explained that this press produces the 16" transcription records for the BBC, which are sent to radio stations all over the world for use in broadcasts.

This 1953 photograph shows 68 years-old Mr. Charlie Knight examining the first pressing of the new "Angel" series. Charlie remembered pressing some of the first "Angel" records when he joined the Company 44 years ago in 1909.

Mr. G. Galpin (left) and Pressroom Foreman Mr. A. Young compare the old with the new "Angel" pressings. Pressroom Maintenance Engineer George had completed 39 years service at this time; 62 year old Albert had served 47 years, having joined the Company in 1906.

We are also shown the few remaining shellac 78 rpm presses. The material is pre-heated in a similar way to the vinyl, but on a steam table without the infra-red hood.

The vinyl records are transported to the next stage of the process, inspection and packing. This department, however, is known to one and all as "examination" or, more usually, "exam". The transportation of the records is done by an ingenious overhead conveyor. The "stands" of discs are loaded on the conveyor at various convenient points around the pressroom, 75 on a 7" and 55 on a 12" stand. Shellac 78 rpm records are transported in large wooden boxes, each containing 99 discs, on wheeled trolleys instead of the conveyor.

Having covered some quality problems during our visit to the audio laboratory our guide says that we will not visit quality control. He goes on to explain that records are collected from each press regularly, are played by quality control staff, and any faults discovered are investigated by the supervisors. When approved by quality control, the samples of production that have been checked are passed to the examination department, which is the final part of our visit.

We are now taken to the next part in the production process, "exam", where the records are inspected and packed. We are introduced to Mr. Sibley, the Departmental Foreman, who proceeds to describe the operation.

All the Foremen gathered together in their tea room at Blyth Road in the 1960's. Back Row (left to right) Ernie Ainsworth, John Tagg, Jim Perryman, Frank Carter, Ralph Laurent, Cyril Gadbury, Fred Leppard. Front Row (left to right) Harry Pickett, Bob Lindley, Alf Beasley, Albert Sibley, Bert Smith.

The overhead conveyor from the pressroom comes into the department, and a device called the lowerator disengages the stands of records from it. The records are then inspected and inserted into their inner sleeves by the packing operatives. Next they are sent to the record stores, where they are put into their outer jackets, before distribution to the retailer.

It was all URGENT in those days too!! Product ready for inspection.

The 78 rpm shellac records, however, had to go through another process before being sleeved. Unlike the vinyl records, which are trimmed after pressing, they had to have their edges ground. This entailed spinning the discs on a lathe type machine, and applying emery cloth to them. This, we were told, produced a smooth bevelled edge. The operation looked quite hazardous, as the edges of the discs were quite jagged and sharp before they were ground. The 78s were then moved to a packing station, where they were inserted into their paper bags.

Edge grinding the 78's – Jimmy Dowsett, the Finishing Department Foreman and Lily Brain are supervising the process.

Moving through the department we come to what was called the box-making section. We are told that all of the cardboard boxes used in the factory and warehouse are made here, as well as boxes for EMI's televisions, radios and gramophones.

Finally we reached the section known as record returns, where the unsold and faulty product is returned for sorting and return to the factory, so that the plastic material can be reclaimed.

Mr. Sibley tells us that over 150 women per shift are employed in exam, and for the most part are paid on a piecework basis. We learn that the department is also responsible for its own administration, and supplying production statistics for the whole factory.

In the box making section, the operator makes up lids on the stapling machine.

We are now near the end of our tour and look round for Chris who tells us that he is wear testing material formulations for 78 rpm discs. The sight that meets our eyes is almost unbelievable, as six records are being tested simultaneously, and, as with the LP wear test, three heads are playing each disc. When the three heads reach the lock groove Chris puts a cross on the label, changes each needle and puts each head back at the start. He does this six times for each disc. There is just time for him to say goodbye and offer us his best wishes before another disc finishes. We thank our guide and make a mental note to write and thank all those involved in our visit. We remember to pick up our own record and walk out into Blyth Road after a most enlightening visit.

THE NEW SITE

A top secret 'special project' was established at the highest levels in the late 1960s to relocate the record factory to a new site. This was designed to provide solutions to several problems, which were stated, at the time, as follows:

> No economic expansion, capable of meeting the expected continuing increase in the demand for records, was possible on the Blyth Road site.

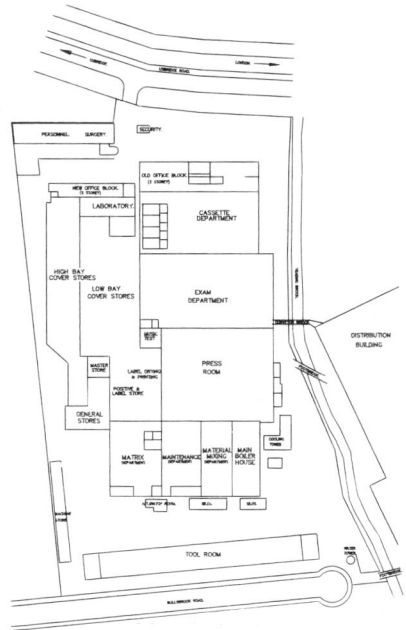

Production Site – EMI Records (circa 1975).

The possible changes in the nature of the business, e.g. the move from disc to eight-track cartridge or cassette, and the possible introduction of video discs, must be catered for, and there was not enough room at Blyth Road.

Seven separate buildings, with a mixture of single and multi-storey types are currently in use at Hayes, which limits efficiency and does not allow a modern flow pattern lay-out to be established.

The Hayes multiple business site, involving Electronics and Tape as well as Records, imposes serious labour problems that are well known to the EMI Board.

Various sites were investigated and one in Stevenage was considered suitable for the needs of the business, but was later rejected on the grounds of cost and the formidable labour problems, including major redundancies, which would have arisen.

So it was in July, 1969, that approval in principle was given by the Parent Board for moving the U.K. records manufacturing operations to a new site on the Uxbridge Road, in Hayes. At this late stage, the opportunity to purchase land at the rear of Dawley 1 & 2 buildings was considered. It did not find favour with members of the Board. This was primarily because it failed to overcome the labour problems associated with retaining the record factory on the Blyth Road site.

The selection of the Uxbridge Road site as the new manufacturing facility for disc and tapes became general knowledge when the following 'notice' to staff was published:

The Gramophone Company Limited (EMI Records)
NOTICE TO STAFF

To take full advantage of future opportunities in a highly competitive market, it is essential that we have the facilities to expand the manufacturing and distribution sides of our business.

An aerial view of the Uxbridge Road Site. Construction of the Distribution Building on the Gabriel Site is well underway. Preparations for the new high roof boiler house buildings are progressing.

Our existing premises at Blyth Road cannot be further developed for this purpose; but plans have been developed to secure a highly suitable alternative - namely, an existing single-storey factory building, next to which will be built a custom-designed warehouse, both with office accommodation.

These plans are subject to all the necessary approvals being received, but we wish to take the earliest opportunity of informing our employees of our intentions.

These new premises will form the centre of our future manufacturing and distribution activities. They are located at **1-3 Uxbridge Road, Hayes**, adjacent to the Yeading Brook and some two miles from Blyth Road.

We plan to arrange a number of visits over the next few months to enable representative employees of the Record Company to see the new site, which we anticipate will become fully operational by **1972**.

Further details of our transfer of operations will be made available in due course, but we wish to stress that we have deliberately chosen a site in the Hayes area to facilitate transfers of employees to the new premises.

The move to Uxbridge Road is part of our programme of planned expansion, designed to benefit all concerned, and will be accompanied by a considerable improvement in physical working conditions. We hope that through joint consultation we can rely on your continued co-operation and goodwill to ensure that the transfer takes place as smoothly as possible.

August 1970   **EMI Records**

Owned by Brixton Estates, the renovated premises had remained unoccupied for some considerable time following a protracted strike by the Wolf Rubber workforce, which resulted in a decision by the Wolf Brothers, the previous tenants, to cease operations on the site. P. J. Cow Limited, manufacturers of Li-Lo products had subsequently occupied the site for a short time.

Once the announcement had been made, there followed a period of intense activity under the leadership of the then General Manager, Roy Matthews. He headed a project team comprising Managers Alan Byrne, Ken Butcher and Ron Turvey, with the resources of the large engineering office available to them.

Production Engineers were responsible for designing departmental layouts with Colin Brown covering the pressroom, Chris Adams the matrix production, label and matrix stores and the label printing department including the darkroom. Frank Bolger was responsible for the material mixing and scrap recovery area, complete with the silos. John Byfield was responsible for both record examination department and cover stores. Cyril Best devised the new toolroom layout. He used models of the lathes, mills, drills etc., on a scaled down floor plan. Cyril's hobby was reputed to be Green Shield Stamp collecting, for he spent all his time at work sticking bits of paper on to drawings, as he devised various layout options!

Taking a break from overseas factory drawings, his previous occupation, Layout Draughtsman Ted Riley was kept busy on new factory layouts and some model making. This included the making of a model of the complex under-floor piping arrangement for the pressroom. This was created using coloured plastic tubing of the appropriate scale, and was needed to show detailed requirements to the various contractors recruited to undertake the full size task!

On the electrical side, John Combes, then the Electrical Engineer, dealt mainly with high tension and medium voltage installations, while Peter Williams, the Electrical Draughtsman, concentrated his efforts on lighting and power services layouts. Laurie Stewart fulfilled a similar role on mechanical services and other layouts.

At this time the drawing office, under Harry Hutton and employing ten draughtsmen, was more than fully employed on work associated with major projects of the time. As well as the new site, these projects included both the new 7" injection moulding machines and the new 1400 automatic press, involving Engineers John Simmons and Alan Thompson. It is perhaps worth mentioning all the others also employed in the drawing office at this time.

They included Fred Tanner and the two Johns [Pemberton and Ratcliffe] who were, with Frank Bolger, transferred from the main Dawley toolroom to the record factory engineering office in 1966, when it was further expanding to cater for a programme of equipment development.

Also present was Geoff Pullen who was devoting his efforts to drawing details of the new 1400 press, a duty shared with John White, who perhaps is best remembered for inflicting some considerable damage to Harry Hutton's personal and private car!

The first silo base is located in position and some boiler house equipment is visible. In the background the distribution building has yet to be cladded.

A junior office member at this time was Bob Kew, who was to take up an appointment with EMI Australia in the early 1980s.

An experienced draughtsman, who had strong connections with the Brush Electrical Company in earlier years was C. A. Brown; between frequent puffs of smoke from his smouldering pipe, Alec, as he preferred to be known, produced layouts by the score. The compounding department with its extract fans and lines, silos, dry blend lines and the like fully occupied his efforts in the twilight of his career. The team was completed by Martin Sharman who, during his eighteen months of service in the drawing office, fulfilled the role of Junior Draughtsman.

Insurance requirements demanded the building of several brick built fire walls. The large area destined to be the pressroom was in a constant cloud of dust for many weeks, as the trenches were excavated to hold the under-floor piping, using pneumatic drills and mechanical diggers.

The Pressroom pipework is at an advanced stage of completion. Trunking for air conditioning systems awaits installation.

Contracts were awarded to Energy Equipment as design consultants for the power house, and Weldtite undertook the pressroom pipe-work installations. Industrial Contract Services featured prominently in the preparation of the site, and the electricity board also played a significant part.

The provision of reception areas in both manufacturing and distribution and the attendant switchboards led to Doreen Duckett being recruited from Blyth Road to be chief receptionist. Heading a small team of attractive ladies, she came equipped with much professionalism, which was appreciated by all.

Twenty-six Windsor SP130 injection moulding machines were delivered direct to the new site from the manufacturers from mid-1971 onwards, and positioned in their respective pressroom locations. Toolroom Chargehand Harry Baxter and his team, consisting of Freddie Freshwater and Alan Howard, were to equip these machines with the EMI designed label loading, record transfer and trimming assemblies. As things progressed, they were joined by others from the toolroom at Blyth Road to complete both pneumatic and other pipework installations.

The new 7" presses ready for production.

The presses were commissioned by John Simmons in readiness for the phased transfer of production personnel to the new site from April 1972 onwards. Alan Thompson fulfilled a similar role on the 1400 presses when quantity construction commenced in the toolroom later in 1972.

Elsewhere the site was progressively transformed as air conditioning was installed, new storage racking built and overhead conveyor systems took shape. 12" manual and automatic presses arrived from Blyth Road and were quickly installed. Plating baths and ancillary equipment for the matrix department simultaneously arrived. Power services were up and running and, by early March 1972, the first 12" record was pressed in the early hours of one morning, having burnt the midnight oil. This paved the way for Sir Ian Jacob to press the first official record when The Gramophone Company Board made their site visit on 15th March 1972.

The timetable for the move involved the record stores, or distribution as it became known, undertaking the colossal task of moving over six million records during one weekend in June, 1972. The transfer of production from the old factory to the new, however, was a more gradual process. In fact it had already started. By this

A Carter PS6 air conditioning unit installed in the Pressroom.

time several new automatic, and existing manual presses, were already installed in the Uxbridge Road pressroom!

The movement of Blyth Road presses and the installation of the new presses at Uxbridge Road was phased in this way over a period of several months. Production continued at both sites until the new pressroom could cope completely with the total pressing requirements of all labels, both EMI's own and licensed.

The appropriately named in-house newspaper "EMINEWS" issued a special supplement explaining the reasons for, and benefits of, EMI's new production and distribution centre in June 1972,

The [then] Managing Director, Philip Brodie, explained the reasons for the move by stating that the famous Hayes factory was over 60 years old and spread over seven different buildings, with consequential product flow problems. "The new premises are keyed to the 1980s rather than 1908 and we now have two rectangular buildings directly linked by a conveyor." The new factory offered the opportunity to expand total production, which was becoming harder on the old site as there was no further room for expansion.

In response to the question "what will be the benefit of this expansion to EMI?", Philip Brodie answered: "We shall have the facility to make better records. We shall be putting in new machinery, and handling problems in the new layout should be avoided. We are taking the opportunity to air condition both the pressroom and matrix departments to provide clean air and humidity at a constant temperature. This will result in even higher standards of quality control than EMI has attained over the years. For the consumer this means even cleaner records made under constant manufacturing conditions, and there should be an improvement in both the making of the original matrix and in the plastic moulding. The new factory also has a vastly increased capacity for tape duplication. As you know, the market is really taking off now and EMI will remain in the forefront in terms of both quality and volume."

When asked what improvements the staff would notice in the new site, he replied "It will provide a better environment for all our people. We have consulted with them and their representatives from the very beginning. We had a meeting with union representatives even before the deeds of the new site were signed. We chose a site only two miles away from our present one in order that we could take with us those people who wanted to remain with the company. The expertise of all our production and distribution people is a very important factor in the quality of the product and maintaining our service. By going to a new development area we could have had a lot of financial help from the Government. We didn't do so because we value very highly the skill and experience of all our present staff."

Philip Brodie praised the efforts of Alan Boxer, Cliff Busby and Roy Matthews for the move going ahead on schedule, and acknowledged the assistance they had received from both the Chief Engineer, Alan Byrne, and the Accountant, Roy Morgan. He also extended his thanks to Jimmy Stephenson for "giving us the benefit of his wisdom and experience".

Production Division Manager, Roy Matthews, elaborated on the increased capacity of the new plant: "The combined new sites [distribution, or "stores", and manufacturing] occupy 16 acres and accommodate 432,000 sq. ft. of floor space, of which production accounts for 203,000 sq.ft. A total of 1,900 people will be employed, of which 1,000 will work in the production areas."

The new sites' recorded tape manufacturing department was about three times the size of the comparable area at Blyth Road, but the real difference was in capacity. It was four times as much as in the old factory, evidence of EMI's faith in the future growth of tape.

Roy then described the fire and security precautions offered by the stout internal dividing walls in both buildings. They were equipped with heavy fire-proof doors, which shut automatically at the temperature of 155 degrees Fahrenheit. The system would also trigger off the sprinkler system near the fire. The heavy walls and doors also provided maximum site security, coupled with a sophisticated and extensive alarm system that could detect an intruder on any part of the site. Roy noted that AFA Minerva, a company within the EMI Group, had supplied both the fire doors and fire protection systems.

He also commented on other unique and new features of the site: "In the matrix area, a special plastic piping has been used to protect the system from the effects of acid, and the flooring is highly acid resistant. Here, too, and in the tape area, there are translucent ceilings, hiding all the overhead air-conditioning pipes and other services."

"Nearby is a master vault, soundly constructed to protect the masters in current use in the factory."

"The view from the rear of the factory is perhaps the best. From here you can see the newly-constructed 65 foot high building, which houses the mixing and services areas. The Serck Visco dust-excluding system adds a pleasing, if somewhat angular, architectural touch to the sub-station, which distributes 3.5 megawatts of electricity, to supply the needs of the whole site."

The Serck Visco extractor unit seen here with 5 x 60 tonnes and 3 x 100 tonnes material silos.

On the question of material flows, Roy observed that "automatic batch mixing [of the dry blend powder used to make the records] similar to that in use at Blyth Road, has been installed but the system in use here will be more sophisticated, more reliable, more accurate and more flexible."

"The factory has 112 record presses [a contemporary layout drawing actually showed 115 presses installed] 16 more than Blyth Road, and room for a further 28. Each of the presses has its own P.V.C. material hopper, not unlike the present system, and a scanner is continually checking these and supplying more P.V.C. powder through overhead ducts."

"The ratio of automatic to hand presses will of course be greater, and the new automatic presses, which have been designed by EMI and built to their specification, are already installed. Scrap material [the flash off the moulding] is dropped from each press on to an underfloor conveyor which transports it to the reclaiming machinery housed in the services' area." [In practice this proved less than successful, and was replaced in 1979 by an underfloor air veying system.]

"In the service area, there are three enormous gas-fired boilers, using natural gas, and these provide steam for heating and pressing. Also housed here are the main electricity panel and all the other services needed in the pressroom; pumps for air-conditioning and for vacuum, water, hydraulic oil and compressed air."

"All the services enter the pressroom along a main duct, from where they split off to service each of the four lanes, which in turn contain a line of presses."

EMI also installed, later, but as part of the original plan, its own gas-engined generator for use during any power failures. It could produce sufficient power to provide the essential services, which in turn guaranteed full production of the factory's most important product at the time of any interruption to power supplies from normal sources. The factory's disc production capacity, incidentally, was five million units a month, compared to Blyth Road's three and a half million.

An overhead conveyor, carrying hanging baskets, passed each press to take the finished records to a roller conveyor and thence to the sleeving area. The press operator completed a docket, as he started a batch of a particular catalogue number, which he slipped into a document facsimile transmitter. The information on it was telegraphed to the cover stores so that records and sleeves could meet simultaneously in the sleeving area [known as "exam"].

Here there were lanes, known as the honeycomb, to take up to 100 catalogue numbers. The records from the pressroom, on their stands, were sorted by catalogue number into these lanes. There was also a further store, for those items where things don't quite go right, for example left-overs of one record from a multi-record set.

As Roy said in summing up: "The new pressroom has many advanced features, underground ducting of services, automatic recovery of record flash, air-conditioning, electronic docket information transmission, and so on. Its vastly improved capacity will be achieved with the help of the new automatic presses, EMI designed and built, and, perhaps most important of all for those who work in it, it will be altogether cleaner and lighter than the pressroom at Blyth Road."

Nothing, however, goes entirely according to plan, and the need to expand the storage area for record sleeves in the manufacturing site soon became obvious. Within two years of the completion of the move to Uxbridge Road, a new record sleeve store was constructed to hold over 10 million standard record sleeves under one roof. It claimed to be the largest record sleeve store in the United Kingdom. Eight lines of racks rising 55 feet above ground level and 200 feet long were serviced by four manually operated cranes that moved along rails, up and down, servicing almost 18,000 locations. The store was equipped with its own computer, which could readily identify the locations of sleeves of the 7,000 titles then in the EMI sales catalogue.

The High Bay cover stores in the foreground with Compounding, Toolroom and Distribution beyond.

The sprinkler system in the new sleeve store, an extension of the main factory's network, was installed at three levels in the building, at roof level and at two lower positions inside the banks of racking. A special feature of the fire protection system was the use of a water sprinkler on top of the roof, which directs water down the outside of the walls. This was to provide a water curtain to cool down the aluminium clad walls of the store in an emergency and to help prevent fire spreading to adjacent buildings.

The record stores, now known as distribution, moved location simultaneously. Cliff Busby, General Sales & Distribution Manager outlined the benefits of the new distribution building at the Uxbridge Road site:

"The main distribution area occupies 176,000 square feet of space between the Yeading Brook and the Grand Union Canal, with 45,000 square feet of offices alongside. At present the distribution volume is 200,000 records a day [during peak selling periods]. That represents 70,000 hours of recorded sound; enough to keep you at your record player for 10 years without sleep!".

At Uxbridge Road the new distribution centre was then capable of handling 300,000 records a day, a 50 per cent increase in the capacity at Blyth Road. Cliff Busby, commenting, said "The move is essential not only to maintain EMI's present service - the best of any record company in Britain - but also to enable it to handle a considerable increase in volume over the next few years. Although we are planning for the future, in five or ten years' time the ultimate object will be the same; to get the records to the retailers as quickly as possible. At the new site, everything is in one building and there is a greatly increased capacity to provide for future expansion."

A TOUR OF THE UXBRIDGE ROAD FACTORY

We have now moved forward to 1990, and are about to be taken on a guided tour around EMI's Hayes manufacturing site on the Uxbridge Road, where it has been established for some 18 years.

As we enter the main gate [known as gate number 2, number 1 being the entrance to the distribution site] we are greeted by John Thomas, the security Sergeant, who registers our visit, and hands us our visitors passes. He then escorts us a short walk across the car park to the main reception area, where we are met by our guide, Mike Brooklyn.

Inside the reception area we notice the wooden trumpet and base of the gramophone taken by Scott on his ill-fated expedition to the South Pole, which is mounted in a glass case, and still in working

Mike Brooklyn meets visitors at the works entrance of the Manufacturing Site.

order. Above the reception area is a giant model of "Nipper" and a gramophone.

Our guide tells us that the product formats currently being made at the Hayes site are 17cm [7"] and 30cm [12"] vinyl records, and music cassettes. As we go back outside, leaving the reception area, he points out the foundation stone at the front of the building. This is the one that was originally laid by Dame Nellie Melba, at the site of the old record factory in Blyth Road, in 1907.

Entering another door, known as the works entrance, Mike explains that the new-looking office block on our right is the personnel and wages department, under the control of his manager, John Munro. The offices to the left, we are informed, house the engineering department, power services engineers and the CAD room. CAD is a computer that has replaced the old-fashioned drawing boards and is used in the design of components for the machines and factory layouts.

We walk along a rather dark passage where, on the right-hand side, the walls are lined with information about the site. There is a picture gallery of the management team set in a frame, followed by charts for each product. These illustrate the number of units produced each month of the current and previous years. On the opposite wall as we move down the passage, decorative displays of record sleeves are arranged, showing some recent chart successes that have been made in the factory.

For a factory there is very little noise at this stage; the whole area is very quiet. After we go past the cloakrooms on our right, we turn right, and walk into the department known as the cover stores. Here all the paper parts for records are stored, such as sleeves, bags, inserts, pamphlets and booklets. We are greeted by and introduced to the Supervisor, David Wright. He explains that the store area we are in is known as the "low" bay, and is used for fast-moving sleeves, etc.

He then directs us to the "high" bay. This is a huge storage facility, having 8 lines of racking, 16 locations high, with cubical cranes set between them, providing access to almost 18,000 locations. It holds approximately 17,000,000 units of paper parts in total. The capacity, we are told, has been expanded since the store was originally constructed, by changing its internal layout. One of the staff, Laura, shows us how the crane works, as she drives one into bay number 2, moving through both the horizontal and vertical planes simultaneously, and retrieves the sleeves that are needed for production.

The storage locations are controlled by a Honeywell computer, which randomly locates components in the store, and informs the production planning department that the items have arrived. When the sleeves are needed for production, it produces tickets for the operators to tell them where they are located within the store.

Leaving the cover store we enter the next department on our tour route, the disc production planning office. We are surprised at the high level of activity, with the telephones seeming to ring incessantly, and people from other departments going back and forth. The guide explains that the Hayes manufacturing site has adopted a strategy of being a custom manufacturer. That is, it produces records not only for EMI, but also for Polygram, Chrysalis, Virgin, BMG [better known as RCA and Arista] and many others. The factory at Hayes now manufactures somewhere in the region of 68% of all vinyl records sold in the U.K. marketplace, as well as some for export.

John Yeoman, the Manager, meets us and introduces some of his staff, including Stella Lewin and Jack, who look after EMI and Chrysalis orders. We also meet Joanne Everett, who controls orders for some of EMI's other customers, and Fred who controls the cover store computer.

The planning system is also facilitated by the Honeywell computer. When an order is received from the customer it is entered onto the computer, which generates a bill of materials for the record concerned. The system checks the paper components required, sleeves and labels, etc., to ensure their availability, and places an order on the matrix department, to produce the metalwork, called "stampers", which are used in pressing the records. When the stampers are produced the computer is updated, and if all the components are available the order can be scheduled to the pressroom for manufacture. We are told that the factory currently has over 17,000 different titles that can be ordered on it at any time!

Leaving the planning office, we continue down the central corridor again, passing on our right the label stores, which hold a stock of some 25,000,000 pairs of record labels. Our attention is drawn to the left of the corridor, where two large brown machines stand; these, we are told, are label-drying ovens. As we watch, rows of record labels placed on rods are passing through a heated air-drying oven. This, it is explained, is necessary to remove any moisture from the labels before pressing the records.

If this was not done, our guide continues, the label face would explode when the steam temperature and hydraulic pressure was applied in the record press. This is because any water in the label would expand as it turned into steam. The labels are then sealed into air-tight packets to prevent moisture from the atmosphere entering them. We are beginning to realise there is more to making records than meets the eye.

Once more in the corridor, we are now passing on our right the matrix store that holds all the metal masters for the manufacturing of the records. We are assured by our guide that the purpose of these will become clear soon. The next area we pass is the general stores, which holds stocks of spare parts for the factory and its machines. It seems very quiet and orderly, and has a board above its counter showing how successful it has been in meeting the maintenance staff's requests for stock items.

At the end of the corridor is the entrance to the matrix department, where we are met by Robin Allen, the Matrix Department Manager.

He proceeds to show us round this complex area, and explains how the department receives lacquers or D.M.M. copper cuts from the customer's studios, and then introduces us to Jackie Hinge, the department's administrator. The lacquer, we are told, is cut at recording studios, on a lathe. The cutter on the lathe responds to the magnetic impulses of the original tape recording of the music, and the shape of the cut groove is unique to the sound on the tape.

Jackie shows the method used to log the receipt of these lacquers, by catalogue number, on a PC programme. [The old practice of having differing catalogue and matrix numbers has now ceased.] The computer then produces an instruction to the department to manufacture a "master" and a "positive".

Once logged in by Jackie the lacquer is cleaned and then sprayed with a silver nitrate solution. This is quite dramatic to watch, as the lacquer goes from being a shiny black colour to having a mirror-like silver appearance. We are told that the depth of silver deposited is about 3 milli-microns. This silver plate is essential for the next stage of the process.

The silvered lacquer is mounted on the rotating turntable of the vertical spindle plating unit, and immersed in the plating solution. Also in the plating bath are pure nickel ingots. A current is struck between the pure nickel and the silvered lacquer, and pure nickel is plated [or deposited] on to the lacquer.

The lacquer is left in the bath until the right amount of nickel is attracted to it [or deposited]. The right thickness, we are told, is about 25 thousandths of an inch. It is then removed from the bath and the nickel and the lacquer are trimmed and separated. The nickel plate is called a negative. This negative has ridges on it instead of grooves.

The negative is placed back into the bath, and the process is repeated. This time, however, the nickel plate that is separated from it is called a positive, and has grooves like a record, rather than ridges like the negative.

The positive is played and tested by matrix quality control, and is then put back into the bath to make stampers, which are used to mould the records. [This

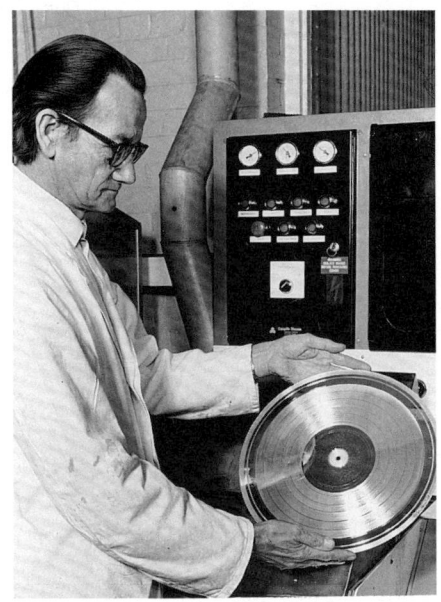

Ron Head examines the silver sprayed lacquer.

The silvered lacquer is prepared for growing in a vertical spindle plating unit.

process has remained almost unchanged in the thirty years that have elapsed since our "tour" of the Blyth Road factory.]

Some customers like to have sample records made before they release them to the public. In this case the first two stampers, A and B sides, are supplied to the pressroom to manufacture "White Label" records, which are sent to the customer for their approval. [The practice of numbering by use of GRAMOPHLTD has also ceased.]

The process is completed by the stampers being booked on to the Honeywell computer as finished, which, we are informed, notifies the planning office, who are then able to release the order for manufacture.

In the event of the order being a 7" record with a painted centre, rather than paper labels, the centre of the positive needs to be etched with the title information. To facilitate this a film must be produced, and Robin takes us to meet Roger Aldworth in the photographic darkroom. Here we are shown how a negative film is constructed and produced, identifying all of the label information, i.e., title, catalogue number and copyright.

We then move into the etch room, where the negative film is used in the etching process. The positive to be etched is dipped in photo-resistant lacquer, and then dried. Next the negative film is fixed to the

Roger Aldworth in the Dark Room.

centre of it, and placed into a unit which exposes the coating to ultra violet light.

The positive is then moved to a machine which develops [photographically speaking] the label copy which has been exposed to the ultra-violet light. Our guide now explains that the positive is ready to be etched, but before this can be done the music grooves are protected with a covering of blue plastic film.

Now the positive is placed in the etching unit, where a solution of ferric chloride dissolves the nickel in the areas developed in the photo-resistant lacquer. When this has been completed the plastic film and photo-resistant lacquer are removed, to reveal the label information etched into the positive.

It is not surprising in such a complex process that a chemist, Tim Stephens, is employed to ensure the process is continually monitored and controlled. He answers a few technical questions from some visitors in our group.

It has taken some time to reach this point and we are offered a welcome cup of refreshment in the comfort of the "rest room". Here the occupants are friendly and polite, and seem to enjoy having visitors to see what they do.

Tim Stephens with a special disc in the Etching Room.

We leave the matrix rest room, and pass on our left the maintenance department, where machines are taken for major overhauls. Usually, as now, most of the staff are out in the factory doing repairs and maintenance "in situ". We are again outside the building, but this time at the rear of it. There is in front of us a long building that we are told is the toolroom. In here much highly skilled machine work, including milling, grinding and turning is undertaken by the 11 men now employed here. They manufacture spare parts for much of the factory's equipment. Mike informs us that the member of staff in the toolroom with the shortest company service has been here nearly 30 years!

We leave this building and return to the main factory through another door and, looking up as we do so, we can see a total of eight massive silos. We are informed that these contain both the raw P.V.C. and the mixed material for making the records. As we enter this vast area, we are met by the Supervisor, Dennis Bendall, who is introduced by our guide. Dennis fascinates us with an explanation of how he single-handedly mixes the base polymer with "carbon", to make it black, "wax" to add resilience, and lubricants and stabilisers.

The mixing is accomplished in large vats, into which the constituent ingredients are automatically blown from their various silos. Paddles then rotate in the vats, mixing the ingredients for about 4 minutes. The mix is then automatically fed into other silos, ready

Dennis Bendall attends the panel which controls automatic weighing and material feed.

for use. We are then shown the "panel". This is an electrical control board for the material process, that also shows it visually, from the receipt of the material, its storage, mixing, transfer to the silos and, ultimately, the feeding of every press automatically. The panel is about thirty feet long and, with its myriads of multi-coloured lights, seems to resemble the kind of control panel seen on the Starship Enterprise! To say it is impressive would be an understatement, and Dennis oozes pride in his responsibility.

Our appetite whetted we now enter the pressroom; are we going to see the records made now, we wonder? No, we turn right along the wall and then right again and enter the power house, a cacophony of sound. In here are three boilers, and all of the compressed air, vacuum and hydraulic requirements of the entire site are generated within its walls. We are introduced to the power services supervisor, Terry Conroy, who shows us round, although the noise makes hearing what he says somewhat difficult. We are also shown the enormous cooling towers used to control the water temperature, which are situated outside the building.

Leaving the power house, we continue to our next port of call, the pressroom, where the Manager, Mike Russell, takes us first to the office area and introduces the Pressroom Controller, Stan Poole. Stan explains that this office is where the component parts, such as inner bags, labels and stampers, for customers' orders are collated and the order allocated to a specific press.

Injection moulding press making 7 inch records.

There are 59 12" presses and 34 7" presses in the pressroom, able to produce over 1,500,000 records per week. Each press is monitored on a system called Dextralog and the screen shows what customer order is on which press, how many records have been made so far, whether the press is operating properly, and if not, why not!

Mike then takes us back to the pressroom where we see the actual records being pressed. The presses are almost fully automated.

We looked first at the 12" or, as they are also known, the 1400 type presses. The dry-blend material from the material mixing department is being fed automatically into small hoppers on each press. These in turn feed the material into steam-heated chambers which cause the material to become plasticised. The machine then extrudes a specified amount of material into a cup, the labels are located on to this, and it forms what is called a "shot".

This shot is automatically placed between the pressing moulds where the stampers are located, one side of the record on top, and one underneath. The moulds close under 100 tonnes of pressure and

1400 type presses at work in Row 3 – note the pressure gauges and ammeter. Using this type of layout enabled 76 of these presses to be accommodated.

steam passing through the moulds raises the temperature to 170 degrees Centigrade. The moulds are then cooled by having water pumped through them. This, in turn, cools the record to about 40 degrees.

The moulds then open and the record is retained in grippers, which hold the flash of plastic around its edge, and move it to a turntable inside the press. Another shot is fed into the press simultaneously, and this is being heated, pressed and cooled as the one on the turntable is being rotated through a knife which trims the flash off the edge. Next the record moves to another location inside the press where it slides into its bag, and into a waiting plastic box for temporary storage. The press produces a record approximately every 20 seconds.

The Operator, Bill Goodhew, shows how he operates the machine and deals with the finished records. If they do not need to have a jacket, because they are 12" singles, he puts them into cardboard boxes containing 35 records, and then into the crate that will be sent to the customer. If they are albums, or need further packing, they are sent into another department in their plastic box for this to be carried out. While we have been looking at the press, two other operators, Mike Murphy and Dennis Dunne, show interest in our visit and invite us to observe a similar process on the 7" direct inject machines.

These, they explain, are "state of the art" machines and produce a 7" record every 10 seconds. In these machines the vinyl is also fed from the material mixing department to the press via its own small hopper. These machines, however, use a different grade of plastic, which is heated electrically in a barrel, and then injected to fill the gap between the two stampers. Then, as the press opens, the record is removed from between the stampers, by a robotic arm. This then transfers the disc to the inking station.

Here the label area of the disc is coated with silver ink, leaving the etched title information as black plastic, which is easily readable. The record then travels along to a bagging station where it is dropped into its bag, and collated at the end of the press. The operator can then count them into lots of fifty records, for boxing up and onward transmission to the customer.

Normally an operator would work three 12" or three 7" machines at a time, and they are also responsible for checking the visual quality of the records they make.

Are there any other checks made on the quality of the records we ask? Yes we are told, and are then ushered into the music test department, where 6 people are stationed. One of them, Andrea Brooks, explains the procedure for testing records to us.

Rosemary Kwan-Tat checks the cross-overs and looks pleased with the result.

First we go to a machine known as a DDD, an affectionate term for the "Mark V Disc Defect Detector". This machine plays the record backwards at 78 rpm, from the centre outwards. As it operates it tests electronically for any groove defects, and produces a print-out of the faults it finds, if any, and where they are located on the record.

Next the record is placed on a normal turntable and the operator checks the run-in run-out, track cross-overs and a selection of the music. If the record is passed, the operator signs a book to confirm the check; if it is not passed it is referred to the supervisor. Mike takes us to meet the supervisor, Carol Whitmarsh. She shows how she tests the record, and if necessary checks it under a microscope to identify the cause of the problem. Seeing a record groove magnified 40 times really proves how precisely moulded records have to be!

If Carol confirms that there is a fault she reports it to the pressroom supervisor who will take corrective action, including checking back through earlier production until the first occurrence of the fault is found. All the defective records are destroyed, and the vinyl reused.

We are told that EMI's customers return less than 0.1 per cent of the product they make, so the checks would seem to be quite effective.

We leave the testers and walk over to the finishing department, [formerly known as exam] and are introduced to Mel Veness, the Finishing Department Manager, and Tess Gannon, the Supervisor. The records arrive here in plastic boxes, on trolleys from the pressroom. The sleeves and any other parts arrive from the cover stores. They are then matched together and given to the operators to pack.

There are some 16 ladies packing record albums into their outer jackets, and adding posters or applying stickers if necessary. They then put the finished product into cardboard boxes, which are then placed into cages ready for despatch. The lady who is showing us how the records are packed, Winnie Williams, achieves a remarkable speed, and tells us that records are still packed by hand because EMI couldn't get a machine to do it as fast!

As we prepare to leave the vinyl operation, and reflect on what we have seen, a sense of the historic importance of vinyl records to recorded music is apparent. It has brought music ownership to millions, and although its day may be over, it will still be remembered as the pioneer of recorded music for the home.

Mike now guides us into the music cassette area, where we first meet Geoff Grimmel, the Factory Manager, and Bob Bailey the department's Production Manager. They expound with some pride the fact that music cassettes have now overtaken the LP in terms of sales per release. They have become [for what was to prove a short time] the leading music carrier.

Malcolm Goody, the Cassette Planning Manager, is introduced. He shows how an order is entered onto the system. This then builds a bill of materials for it, identifies what components are available and finally releases the order to be manufactured, by way of a job card.

This card is matched up with a one-inch running master tape and passed on to the shop floor supervisor for manufacturing. We are taken to see this tape made in the transfer studios. The Manager here, John Cox, explains that customers send, direct from the recording studios, a digital audio tape [DAT] master, from which the running master will be copied.

He then ushers us into one of the four studios, where an engineer is about to create a one-inch master, and shows us the procedure for loading a one-inch blank tape onto the Studer mastering machine. He laces the tape through its path on the machine and on to its take-up hub. The machine maintains a constant tension and speed control across the recording head and guides. When the recording process is started, the input signal from the DAT is sent to the record head on the

EMI transfer desk type TG 12410. These desks were first introduced into tape records in 1975, and went on to be used worldwide.

Studer via equalising equipment, designed to ensure the accurate recording of all frequencies. John demonstrates that the whole process can be remotely controlled using a separate control panel in the transfer desk.

Having thanked John we move on to the production area, where we are met by the Supervisors, Mary Phillips and Min Emo. Starting at the duplication area we are shown the running master being loaded into the master bin. The one-inch tape is laced through the tape path by the operator, Pat Hood. She sets

Electrosound 8000 series loop bin. These machines replaced the original horizontal bins.

the machine running, which spools the tape into the loop bin at the bottom of the machine. When the end of the tape is reached, the machine stops itself, and the operator then shows us how the two ends of the tape are joined to form a continuous loop.

The tape is then laced through the tape path again only this time across the replay heads of the machine. The machine is then switched on, and the tape flow is stabilised by using vacuum in the loop bin. The tape runs continuously, and the end of each programme is signified by a cue tone, created by two metallised tape strips placed across the start and finish of the tape when it was spliced into a loop.

We are now taken to the duplication machines, known as slaves, and the operator shows us how blank tape is loaded onto them. This tape comes in a form known as pancakes, which are spools of tape approximately 14,400 feet long! The tape is loaded across the record heads on each slave and on to a take-up hub. The music is transmitted from the replay heads of the one-inch master machine to the record heads of the slaves. This is done via co-axial cables to each duplication machine's amplifier, which ensures that the correct signal presentation to the heads is achieved.

It is explained that this recording takes place at eighty times the normal playing speed, 150 inches per second! A pancake takes about

Gauss 1200 series duplication machines. Some of these machines started life in production in 1968. They are better than ever!

19 minutes to record. The number of pancakes needed to complete an order is pre-calculated, as it will depend on how much music is on an individual catalogue number. Each pancake of tape is marked to show its contents. Operators work 10 slaves and one master each.

We are told that one pancake in three is tested from each slave, and, in separate quality control studios, we are shown the method used. It is done by comparing the recorded pancake to a DAT copy of the original studio-produced DAT. Checks are carried out for music content, cue tone spacing, fade and Db levels. Once the quality control operator has passed the pancake the production run can proceed to the next stage, cutting or winding.

In this area of the cassette factory we are shown how the machines work by an operator, Maggie Shah. Two pancakes of tape are loaded on the machine, and the tape has been threaded through to a vacuum block. The tape length is set into the machine, and cassettes with only the pink leader tape inside, known as C-Zeros, are loaded into their magazine.

The machine is then started, and the leader tape from the first C-Zero is pulled out by an arm on the machine, cut, and spliced on to the recorded tape from the pancake. The cassette case drive is then engaged, which winds the programme into the cassette. At the end of the programme the cassette case drive stops, the recorded tape is cut, and then spliced onto the remaining piece of pink leader tape from the C-Zero.

This cassette is then taken away, a new one inserted and the whole operation starts again. What amazes us all is that it only takes 10 seconds to load the tape into the cassette, despite the complexity of the machine's task!

The loaded cassettes are placed in trays of 100, and passed to the final stage of the process, printing and packing.

In this area the cassettes are passed through a series of machines that extend for about 30 feet. The first machine is a printer, and has two printing plates mounted inside it, which transfer the title information on to the cassette. The cassette is then taken away by conveyor to the packing machine, which inserts both it and its inlay into the library box.

From here another conveyor takes it to a machine that will put a sticker on to the library case, if required, and from there it is transported into an over-wrapping machine. This covers the cassette with a clear foil wrapper, and sends it to the final machine in the line, which bulk packs the cassettes into twenties for onward transmission to the customer. This line of machines works at an incredible rate of greater than 3,600 per hour! We can see that there are three such lines in the factory.

Our guide directs us now to the last department on our journey, the despatch area. Articulated trailers are lined up, with their tail gates open, crammed full with finished product ready to be sent to the customers.

John Eddington is completing the despatch manifesto of a shipment to the Polygram distribution centre, while Jim McKinley, with Malcolm Bennett, prepare a delivery to the EMI distribution centre.

As we conclude our tour our guide takes the opportunity to introduce Peter Hall, the Manufacturing Director. He courteously hopes we have enjoyed out visit and accompanies us to the gate where our passes are retrieved and a fascinating visit to the world of manufacturing entertainment is concluded.

OVERSEAS FACTORIES

From its earliest days the company had been an international one, with its headquarters in the United Kingdom. While this book is focused primarily on the Hayes manufacturing sites, engineers, draughtsmen and others from the Hayes factories have been responsible for the design and construction of many of EMI's factories around the world. It was felt that it might be useful to look at some of these factories to give an idea of just how widespread the influence of the Hayes staff has been.

The Gramophone Company was formed in April 1898 by Trevor Williams as a small private trading syndicate of family and friends. Trevor Williams was a 38-year-old Lincoln's Inn solicitor who was to guide the Company for the next thirty-two years.

The new Company took part of a building at 31 Maiden Lane, close to London's west-end theatres and music halls. Here the first recordings were made early in August 1899 on equipment set up by a young American, Fred Gaisberg. A year later The Gramophone Company Limited was formed with Trevor Williams as its first Chairman.

The Company was re-registered as the Gramophone & Typewriter Co. Ltd. in December 1900, following acquisition of the patents for the Lambert Typewriter. The famous 'His Master's Voice' picture was also registered as a trademark in Britain at this time.

New larger head office premises at 21 City Road were acquired in February 1902, and the Company letter heading highlighted its international nature by listing branches in:

Berlin	Moscow	Milan
Hanover	Brussels	Barcelona

Paris	Amsterdam	Lisbon
Vienna	Stockholm	Sydney
St. Petersburg	Copenhagen	Calcutta
Cape Town.		

The Gramophone Company's first pressing plant was opened in 1898 by its German subsidiary in Hanover, and until 1907 most of its records were manufactured there. Such was the success of the gramophone record that by 1907 it was necessary to build a factory in India. In the same year a large tract of land was acquired in Hayes, Middlesex, upon which the new works for the Gramophone and Typewriter Co. Ltd. were to be established. This factory was not, however, the first in the U.K. to manufacture records, as The Guinness Book of Recorded Sound states that the first pressing plant was set up by Nicole Freres in 1903.

The Company reverted to the title The Gramophone Company Ltd. in November 1907 following cessation of typewriter manufacture, sales and service as from 1st February 1905. This title was to continue in use until it was replaced by EMI Records Ltd., from 1st July 1973.

No doubt the First World War of 1914-1918 delayed further expansion plans, and, indeed, the Russian branches were lost, but

The Gramophone Company of India Limited factory at Dum Dum, Calcutta.

The C.I.E. horse drawn cart awaits records outside the factory at Waterford, Eire.

many new companies with manufacturing facilities were established in the 1920s. Shanghai, in China, commenced manufacturing in 1922 and many others followed, including Argentina and Australia in 1925, New Zealand in 1926, and Greece and Turkey in 1929.

Some factories were, however, on a very modest scale, and some were extremely unusual in their operational methods. For example, EMI became established at Waterford, in Southern Ireland, in 1936 and manufactured product for all the major record companies represented there [including Decca and Philips as well as EMI]. This was done at its small plant, with four 7" presses, later supplemented by one 12" press. The facilities were accommodated within buildings occupied by The Tyresole Rubber Company whose business was remoulding motor vehicle tyres. Their processes necessitated services, for example high pressure steam and water, similar to those required for moulding records.

The quality of the records was compromised, however when remoulded tyres were in high demand. This caused steam pressure to fall below an acceptable level for record moulding! In the late 1950s this record factory remained unique, in as much as all its product commenced its distribution journey by horse-drawn cart to the C.I.E. railway station at Waterford. A new plant, equipped with modern 7" and 12" injection/compression presses, was opened in Dublin in the mid 1960s and resulted in the closure of the Waterford plant.

The Second World War of 1939-1945 resulted in the loss to the company of several installations, including those in Shanghai, Berlin and Warsaw.

In countries where The Gramophone Company did not have its own manufacturing facilities, the market was served either by importing product from elsewhere or, often, by local manufacturing licensees.

In Brazil the manufacturing facility was at Sao Paulo, this photograph displays the previous company title.

In Scandinavia, Heger Plastics were licensed to produce records at their factories in Oslo, Norway and Stallarholmen in Sweden, alongside their own products, which were generally injection moulded components for the medical world.

Similar arrangements existed elsewhere, particularly in South America, with licensees producing records in Columbia, Venezuela, Ecuador and Peru. These complemented EMI companies with manufacturing facilities in Argentina, Brazil and Chile.

With the introduction of vinyl records from the early 1950s onwards additional manufacturing plants were opened; this was supplemented by the acquisition of majority/controlling interests in other record companies. A major acquisition was that of Capitol Industries Inc. in 1955, one of the largest and most powerful record organisations in the United States, and throughout the world. Founded in 1942, the company operated three large pressing plants in the U.S.A., with its headquarters in Hollywood's Capitol Tower and executive offices in New York.

In 1931 The Gramophone Company had established large manufacturing premises at Kawasaki in Japan. Kawasaki was 12 miles from Tokyo and 6 miles from Yokohama, and the plant was equipped with thirty-four 10", and six 12" presses. The company operated as the Nipponophone Co., and yielded annual outputs exceeding six million records during the period between 1931 and 1935. The fate of this factory is unknown after 1935. Capitol-EMI were later successful in obtaining an investment in Toshiba Musical

Industries, a subsidiary of Tokyo Shibaura Electric Company; it was apparently an unusual occurrence for foreigners to have an investment in a Japanese record company. Toshiba-EMI, or TO-EMI as it became known, has manufacturing facilities in Gotemba, although both its vinyl facilities are now closed it continues to manufacture cassettes, CDs, laser discs and music video.

In the mid 1960s a new factory was built at Caronno Pertusella, for EMI Italiana; a new company formed from the amalgamation of the existing manufacturing and marketing companies. The origins of these companies date from 1904, when the Fonotipia Company was formed. This company later became a member of the EMI group and by 1931 was equipped with twenty four standard Columbia presses, and two 6" and one 16" versions. With these it achieved total annual outputs of over one million units.

In 1967 EMI obtained a major shareholding in the Dutch company Bovema N.V. and three years later moved the recording, manufacturing, sales and distribution facilities into new and modern premises in Haarlem.

To provide for continued expansion, a new factory was built for Compania de Gramofono-Odeon, Barcelona in 1969, which also uniquely made provision for the manufacture in-house of record sleeves. The history of this company dates from 1906 when it commenced operations in a small factory in Barcelona. It subsequently transferred to larger premises, again in Barcelona, in 1926, being equipped with 15 presses yielding up to three-quarters of a million records per year.

The opportunity for EMI product to return to the Russian and eastern European market occurred because of a licensing deal coupled with technical support which EMI concluded with Jugoton Records, of Zagreb in Yugoslavia in 1964. Technical staff from Hayes who visited this factory were encouraged by their hosts to down a generous measure of sljivovica [a very potent plum brandy] "to protect their throats from infections"! It never tasted quite the same when drunk elsewhere!

By the late 1960s, the number of worldwide manufacturing facilities was at its peak, with factories at the following locations:

Country	Company	Location
Argentina	Industrias Electricas y Musicales Odeon	Buenos Aires
Australia	EMI [Australia] Ltd.	Sydney
Brazil	Industrias Electricas y Musicales Fabrica Odeon SA.	Sao Paulo
Canada	Capitol Records [Canada] Ltd.	Toronto

Country	Company	Location
Chile	Industrias Electricas y Musicales Odeon SA.	Santiago
Denmark	Electric & Musical Industries [Dansk-Engelsk] A/S	Copenhagen
France	Pathe Marconi	Chatou, Paris
Germany	EMI Electrola GmbH	Cologne
Greece	Columbia Gramophone Co. of Greece Ltd.	Athens
Holland	Bovema EMI	Haarlem
India	The Gramophone Co. of India Ltd.	Calcutta
India	The Gramophone Co. of India Ltd.	Bombay
Ireland	EMI [Ireland] Ltd.	Waterford
Italy	EMI Italiana Spa.	Pertusella
Japan	Toshiba Musical Industries Ltd.	Gotemba
Lebanon	EMI Lebanon SAL.	Beirut
Malaysia	EMI [Malaysia] SPN. BHD.	Kuala Lumpar
Mexico	Discos Capitol de Mexico.	Mexico City
New Zealand	EMI [New Zealand] Ltd.	Wellington
Nigeria	EMI [Nigeria] Ltd.	Apapa
Pakistan	The Gramophone Co. of Pakistan Limited.	Karachi
Singapore	EMI [Singapore] Plc. Ltd.	Juron
South Africa	EMI [South Africa] Ltd.	Johannesburg
Spain	Compania del Gramofono-Odeon	Barcelona
Turkey	Gramofon Ltd., Sirketi.	Istanbul
U.K.	The Gramophone Co. Ltd.	Hayes, Middlesex
U.S.A.	Capitol Industries Inc.	Jacksonville
U.S.A.	Capitol Industries Inc.	Scranton
U.S.A.	Capitol Industries Inc.	Los Angeles
U.S.A.	Capitol Industries Inc.	Winchester, Virginia

The availability of the new 12" 1400 type automatic press, in the 1970s, led to the introduction of new plants, and the updating of many others. The Cologne factory, however, equipped itself with its own 7" and 12" tandem autos, in-line units utilising a single W & P extruder. The electronic control system ensured that the extruder delivered compounded material at a rate which provided for continuous operation, i.e., that only one press of the pair is moulding while the other is preparing for the next moulding.

In 1974, EMI purchased an existing operational record factory situated in Alkmaar, the cheese-producing area north-west of Amsterdam. The plant was equipped with twin-cavity presses built by Taunus Ton Tek, but these were soon replaced by a number of 1400 type presses. This factory continued in use, complementing the Haarlem facility, until the new large manufacturing and distribution centre opened at Uden in 1978, when both it and Haarlem were closed.

A centralised manufacturing facility was completed at Amal, Sweden in late 1975 to cater for the Scandinavian market, and resulted in the closure of the existing plant in Copenhagen and the cessation of licensed manufacturing by Heger Plastics in Norway and Sweden. Equipped with sixteen of the type 1400 presses and two 7" injection/compression presses, this plant was not, however, to last long in the ownership of EMI, as it was sold to a management consortium in the early 1980s.

By 1977 the well-established Pathe* Marconi, facility in Paris held thirty six 1400 presses, and among other existing plants updated by the provision of automatic presses were Australia, Greece and Spain.

The continued growth of music cassettes and CDs in the 1980s heralded the decline of vinyl records, bringing with it many plant sales and closures throughout the world.

In the U.S.A., the last Capitol vinyl disc manufacturing facility at Winchester was closed and the small demand for vinyl product was met by contracting out the work to other companies.

In Europe, the new Spanish factory was an early victim, being sold in the early 1980s to licensed manufacturers. Uden, in Holland, ceased vinyl disc manufacturing in the mid 1980s, followed by Chatou, in France in 1990.

The Nigerian plants reflected rapid growth initially at Jos in the 1960s, which was soon replaced by a new larger plant at Apapa. This in turn was superseded by yet another plant at Ikdja, finally being sold to an independent record manufacturing company in the late 1970s. The last small vinyl disc manufacturing facility was established about this time with a few manual presses in Bangkok, Thailand.

Elsewhere, New Zealand ceased manufacturing in 1987 as did the Juron plant in Singapore; Australia followed in 1990.

The Pathe Marconi factory was sited on the outskirts of Paris at Chatou.

Here in the United Kingdom, many other record companies faced the same problem. RCA Records ceased manufacturing in 1981 and in 1987 Polygram closed their plant at Walthamstow and embarked on a manufacturing contract with EMI. This was followed in 1989 by the cessation of vinyl disc manufacturing at CBS, leaving EMI Music Services [U.K.] at their Uxbridge Road site as the last major manufacturer in the United Kingdom.

By 1992, EMI continued disc production at only five locations: in the U.K. at Hayes; in Europe at Cologne, Germany, [whose closure was announced in November, 1992] and in Milan, Italy [closed spring, 1992]; at Dum Dum, Calcutta, India [output now 20,000 units per day], and in Brazil.

CHAPTER TWO

The Equipment

INTRODUCTION

Over the years the equipment used to manufacture vinyl records has changed considerably. This has been caused by two main factors: changes in the product, both in the quality and quantity needed; and the need to improve the efficiency and reliability of the factory.

The need to improve efficiency became more important after the Second World War. Although some development had gone before, little had changed in the equipment used, since the start of manufacturing in 1907.

The drive for mechanisation of the processes began in earnest in the 1950s with the introduction of the microgroove record and its more exacting quality requirements. The need for automation was furthered in the 1960s, when the growing popularity of recorded music created phenomenal increases in the level of demand.

While automating the processes was a continual drive within the company, a reorganisation in the late 1950s resulted in special effort being devoted to it. This was the creation of the development department. The individual music test booths, formerly located on the ground floor at the front of the record factory, were replaced by properly sound-insulated booths constructed alongside the main corridor, next to the pressroom. The vacated area was then established as the development department, being equipped with all press services, and adequate electrical supplies.

Machinery installed here included a small lathe, a drill and a mill. The intention was that larger parts would be manufactured in the well-equipped toolroom. The department was charged with responsibility for the development of the equipment and processes within the factory, but with a special focus on automating labour intensive tasks.

Alf Ashworth, an instrument maker, and John Hassel, an ex-Bakelite factory toolmaker, were the only permanent staff employed in the department. It was Bill Toull, however, who made the department his second home. Seconded from the toolroom, he made major contributions to the development departments success designing several new items of equipment while in the department. These included, particularly, the turntable granule pre-heaters, and machines for processing records in the examination department.

In the mid-sixties, the development department was past its peak, in terms of the number of different developments taking place. Two major projects, which were to have such an impact in future years, were still progressing side by side. These were the 7" injection moulding system, and development of what later was to be known as the 1400 press.

The idea of having such a department was renewed in the 1970s, with the establishment of the record engineering development department at Uxbridge Road. This short-lived section, again inhabited primarily by staff seconded from other areas, was located at the east end of the toolroom building.

THE MATRIX PLANT

Before we can describe and understand the matrix processing equipment it is necessary to describe the groove geometry and the replication sequence of the process.

Groove geometry is as follows;

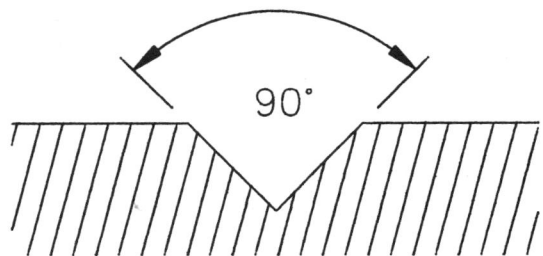

Mono: The groove modulates laterally [sideways] only.

Stereo: The groove modulates both laterally and horizontally [sideways and up and down].

The section of the groove is always the same shape.

Thus the needle on a record player translates the lateral and horizontal movement in the shape of the groove back into sound, in a reversal of the studio's cutting process. In this process the electrically recorded sound is passed through electromagnets that have a cutting tool suspended between them. The change in current brought about by the recorded sound causes the cutter to move sideways and vertically, thus cutting the uniquely shaped groove in the disc rotating beneath it.

The first stage is, as noted above, for the groove to be cut into a cutting medium based on either wax, lacquer coated aluminium, or copper.

Lacquer cut medium.

2nd stage: A master or negative is electroformed from the cut.

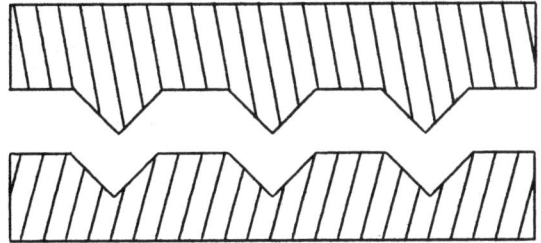

Metal Master

Cut medium

3rd stage: A positive is electroformed from the metal master.

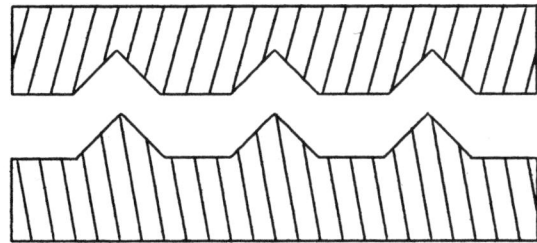

Metal Positive

Metal Master

The last stage is a stamper, the tool from which the record is moulded, electroformed from the positive.

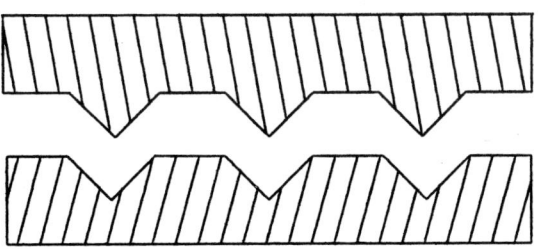

Metal Stamper

Metal Positive

Disc Cutting Engineer, Chris Blair, examines a lacquer disc at EMI's Abbey Road Studios.

Electro forming, or electro plating, can be simply described as an electric current applied to a conductive solution of metal salts, when the positive metallic ions [or particles] migrate from the positive anode to the negative cathode. In our case the anode is the pure metal of nickel, and the cathode is the part to be plated.

Now that the steps in the process and, more importantly, the names of the various parts have been defined, it is possible to review the development of this part of the production process.

Early recordings were made directly through acoustic devices and cut on to solid wax discs. These discs were fourteen inches in diameter and one inch thick, and were cast and machined flat by Jack Manley at Blyth Road. In these days up to six simultaneous recordings were made. The masters were fitted to a press, and test pressings made [in shellac] both for technical approval, and to select the best recording.

During the war years, 1939-1945, wax became very difficult to obtain, and a flow coated wax, developed at Hayes, replaced the cast wax. In this process a special, particularly pure, wax was coated onto an optically flat, 14" diameter glass disc. The thickness of the wax was approximately one-eighth of an inch.

Lacquers [sometimes mis-named acetates] were introduced in the late 1940s. The lacquer is actually a plasticised nitrocellulose lacquer, coated onto an optically flat aluminium substrate. The development of the lacquer disc was pioneered by John Wooler, and it remains in use as a cutting medium today.

The first requirement for the electroforming process is that the cut disc, wax or lacquer is made electro-conductive. To achieve this, the early waxes were covered with graphite powder. This was replaced in the 1930s by bronze dust, which gave a superior finished product.

In the early 1940s the flow coated waxes were made electro-conductive by vacuum gold sputtering, a process developed by Bill Soby and Harry Christmas, at Hayes.

Lacquer disc processing was further developed by Bill Soby and John Harris in the mid-1940s. They developed a technique for chemically reducing silver onto the lacquer surface, a process known as electroless silvering, which continued in use, with automation and further refinements, into the 1990s.

Electroless silvering is now a fully automated process. The unit consists of a metering and microprocessor control panel that governs the sequential spraying of demineralised water, cleaning solution, sensitiser, acid and silvering solution on to the rotating lacquer disc. The solutions are pumped, via a 2 micron filter, to the spray jets. Once the start button has been pressed the process is completed entirely by the machine. When the cycle is finished, the unit switches off automatically, a warning lamp illuminates and a bell sounds.

The next stage of the process is to manufacture a metal master from the electro-conductive cut disc.

The processing equipment for the wax mediums and the early lacquer discs consisted of a rubber lined steel tank containing copper anodes and acid copper sulphate electrolyte. The cathode spindle and chuck were made from ebonite. The plating voltage was 12V DC, and the current density was 50 amps per square foot. The disc to be plated was fixed to the chuck and was rotated, semi-immersed, horizontally, in the electrolyte.

A copper master took 4 hours to grow, after which it was separated from the wax and flash plated with nickel.

The approved nickel-faced copper master was then passivated [cleaned and given a microscopic covering with an oxidising agent]. This enabled separation of the parts to take place after the electroforming process was completed.

Today, however, the equipment used for pure nickel parts is specifically designed to spray a passivating film onto a master or positive before plating. The unit is similar in concept to the automatic silvering unit, allowing chemicals to be sprayed in sequence on to the rotating master or positive. Safeguards included the introduction of detectors on the metering control, which warn if the correct flow rate is not achieved. The early passivating solutions contained potassium dichromate, which oxidised the copper or nickel surface. A colloid solution of egg albumen was used in the 1950s and 1960s, but a universal proprietary passivating agent, with potassium dichromate, is used today.

After passivating, the master was then flash nickel plated, and a copper-backed positive grown.

Bob Ashby prepares a part for filming.

The process was repeated from the positive to grow a stamper. The nickel faced copper stamper was grown to a thickness of approximately 8 thousandths of an inch, and then soldered to a rolled copper backing, a process requiring great care. The stamper was finally chrome plated.

The reason for this apparently convoluted process was that a wax or lacquer can only be processed once, normally. A production run, however, might require many moulding tools to be used, therefore the need to grow the large family tree of electroformed plates!

A simplification of the matrix routine was achieved with the introduction of the nickel based process. In this, nickel anodes replaced the copper, and nickel sulphamate replaced the acid copper sulphate. Again this was developed by Bill Soby and John Harris, at Hayes. It occurred in parallel with the introduction of vinyl records in the 1950s, and by this time direct recordings had been replaced by taped recordings, a significant advancement.

The processing equipment, however, physically remained unchanged until 1959/1960 when the 5 cell vertical spindle plating bath was developed at Hayes by Roy Matthews. The equipment, apart from updated electrical and electronic controls, remains basically unchanged to the present time.

The appearance of the matrix department at Blyth Road was transformed by the installation of the new vertical spindle plating units.

The vertical spindle unit consists of two major assemblies, the electrical supply cabinet and the main frame.

The electrical supply cabinet contains a three phase electrical input, transformer, rectifiers and control circuits. The output is variable from 0 to 24 volts DC.

The main frame houses five plating cells above a 500 litre sump tank. The temperature of the nickel sulphamate electrolyte is controlled by two air operated modulating valves, allowing steam or cooling water to enter the heating or cooling coils situated in the sump tank. The plating solution, nickel sulphamate, is pumped to the five cells, the rate of flow through each cell being manually set. The solution is filtered to 2 microns at the output of the pump.

Each plating cell is fitted with a pneumatically operated opening lid. This is equipped with an electrically rotated turntable, and the plate to be grown from is mounted on to this. The plating current is fed to this plate [the cathode] through a carbon brush fitted to the spindle drive shaft. The anode in each plating cell consists of a titanium basket containing the nickel ingots.

The unit is completely automatic when operational. The part to be plated is attached to the turntable in the lid, and the start button pressed. The lid closes and the turntable rotates to commence the plating process. At the pre-set amp hours required the lid opens and plating stops.

When a lacquer is cut a burr of material on the shoulder of each groove wall, known as horn, is created. The horn is removed at the positive stage by lightly polishing the music surface with mild abrasive. Failure to remove this horn results in the stained appearance of the record surface.

It is necessary for the label area of the 12" stamper to be roughed to prevent label burst at the pressing stage. This was originally achieved by embossing the lacquer with a heated tool. It is now done with a purpose-built tool that is applied to the positive under pressure of 1400 psi.

To prevent damage to the music surface, all 12" stampers are protected by the application of a blue plastic film before the post-growing operations commence.

The optical centring equipment.

The hole in the stamper is produced progressively in the process, and is not necessarily in the centre. In order to centre it accurately, which is necessary to ensure that the record has the music aligned correctly with its centre hole, the stamper needs to be recentred. An optical centring machine was designed by Ray Saunders at Blyth Road in the late 1960s to improve this process. It consisted of a turntable, supplied with vacuum to hold the stamper, and a microscope arrangement allowing the concentric run-out groove to be displayed on a screen. The turntable, mounted on a laterally moveable bearing, is rotated and adjusted until the concentric run-out groove shows on the screen as running true. A punch and die, mounted centrally in the unit, is then actuated and cuts a 1" hole exactly in the centre of the stamper.

The back of a stamper is micro-rough after growing. This roughness would press through to the record, and produce an unacceptable surface noise, and so must be polished smooth. Originally this was done by rotating the stamper at about 360 rpm, and applying abrasive cloth to the stamper back. Now, however, a fully automated machine is used, designed by Paul Waeshler, and made by Capitol in the U.S.A. It was supplied to Hayes in the mid-1980s to coincide with the commencement of the D.M.M. process. The

Ray Houghton with the Capitol back-sanding machine.

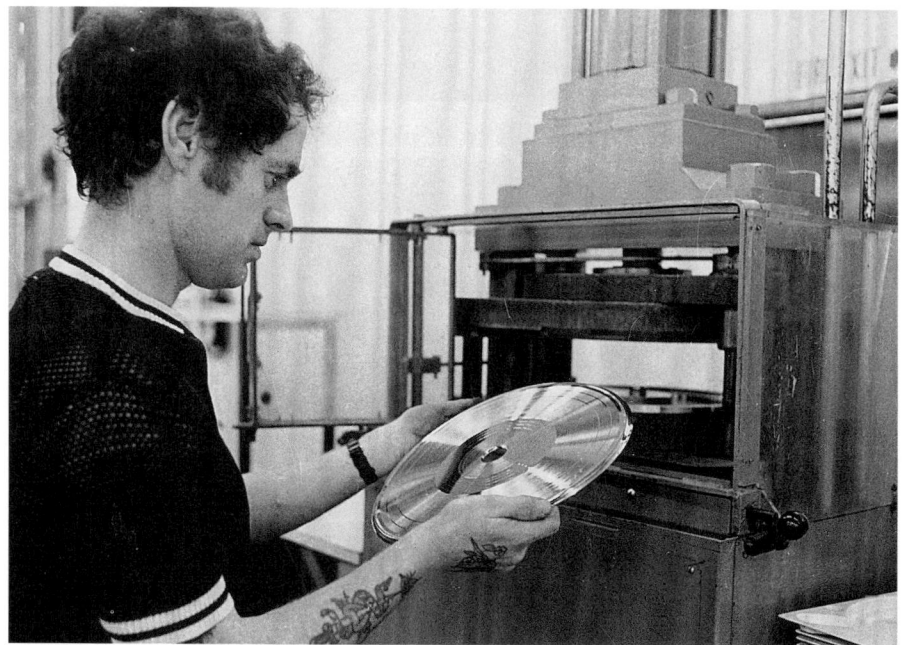

A formed 12" stamper is removed from the hydraulic press.

machine has a motor driven turntable on to which the stamper is fixed, and four air driven motors driving abrasive discs. These sequentially come into contact with the back of the stamper. The abrasives can vary in grit size, and the cycle can be varied infinitely, allowing an optimum polish effect to be achieved.

The stampers were, finally, formed on a hydraulic press at the centre and edge to allow for mechanical clamping in the record press.

The matrix department quality controls all metal parts visually as they are made, and in addition all new positives are 100% aurally checked. It is possible with care, very steady hands, and with the aid of a microscope to repair minor defects on positives!

The matrix department has always been equipped with a laboratory. Its task is to control the plating process through daily checks of plating solution pH, density, nickel chloride content, surface tension and temperature. The laboratory also supplies an analytical service to the site for checking waste water, vinyl materials, paper components and general analytical tasks.

Possibly the final major development in the matrix process occurred in 1982, at an international technical conference, when Direct

Metal Mastering [D.M.M.] was introduced. The D.M.M. process offered a significant technological advance because the music groove was cut into a copper blank, rather than the traditional lacquer. The copper blank was, in fact, copper plated onto a stainless steel substrate.

The D.M.M. process was pioneered by Teldec Records of West Germany in collaboration with EMI Electrola. The technology was purchased under licence by EMI, and introduced to the Uxbridge Road site in 1984. The matrix department, then headed by Chris Adams, aided by Colin Brown and the engineering team, developed and built four inclined plating cells, and a copper pyrophosphate material, to manufacture the copper blanks for sale to cutting studios throughout the U.K., Europe and the U.S.A.

D.M.M. offered advantages not only to the electroforming process, but also to the record company and the consumer, these were:

Longer programme lengths without the risk of jumping or sticking.

No "end of side" distortion.

No need for the copper to be silvered, as the D.M.M. copper disc was conductive.

No need for positive dehorning, as there was no burr produced during cutting.

No post or pre-echo.

The possibility of not requiring a master or positive, as stampers could be grown directly from the D.M.M., time after time.

In the early 1980s a process developed by Philips Records was introduced to Hayes. By chemically etching the label information into the centre of a nickel positive, the stamper and subsequently the record contained the label information. The record centre was run through an inked roller to enhance its appearance.

This was achieved by having the label information produced through a "Compugraphic" computer, and photographed onto a negative film. The positive 7" metal part was dipped in a photo-resistant coating, and exposed to ultra violet radiation. It then passed through a developing solution and finally through an etching solution of ferric chloride. The finished positive was then cleaned and quality control checked to verify the label copy and audio quality.

In the peak production year of 1990, the matrix equipment included:

    1   Silvering unit

   12   5 cell vertical spindle plating baths

- 4 Copper blank plating baths
- 3 Automatic passivating or filming units
- 1 Etching plant
- 2 Centring machines
- 2 Back sanding machines
- 2 Dehorning units
- 1 Stippling unit
- 3 Forming presses

The department was capable of processing over 1,000 lacquers or coppers, and 10,000 stampers, per month.

THE RECORD PRESS

At the time of the introduction of the long playing record at Blyth Road in 1952, the conventional record press had reached its ultimate stage of development. The basic press retained design features dating back to the 1920s, although very few presses were actually that old!

The presses were originally fully manually operated. The press operator loaded his press with the heated, rolled biscuit and labels, between the stampers. Next the counter-weighted press leaf was closed, and the bridge blocks locked under the press by means of a cam handle. Activating a valve that allowed the hydraulically powered ram to rise fully, the operator then applied his experience to decide when to operate the control valve that turned off the steam for heating the moulds, and switched on the cooling water instead. He then judged the right time lapse before simultaneously lowering the ram and turning off the water. The operator then applied steam to reheat the moulds, immediately after removing the moulding from the press, and prepared for the next cycle.

This evocative photograph illustrates an early example of the semi-automatic press in 1928.

Later on, these presses were fitted with motor driven gearboxes, driving camshafts that operated directly on the steam, water and hydraulic valves. Their introduction was a further step toward more uniform quality. They removed the discretionary cycle time created by the operator, and replaced it with one that was mechanically controlled. This type of press, and subsequent versions of it, were known as semi-automatic presses. The mechanical gearbox with camshafts was subsequently replaced by the Bristol timer. This was standard equipment for all Hayes semi-automatic presses from the mid-1950s onwards.

Two distinct types of press existed in both 10" and 12" ram diameter versions. One design utilised a separate well assembly located within the press body, whereas the other utilised an integral casting. The press body was mounted on a separate single or two part cast iron press stand, and in most other respects the two sizes were very similar with only detail differences.

These up-stroking presses used four large springs attached to guide rods, secured to the press bolster, to help the ram downstroke to its lower position. A slipper plate was provided to correctly align the bottom mould with the dowel-located top mould secured directly to the press leaf. A pair of bridge blocks, inter-connected with each other by tie rods and secured to the leaf by passing through bearing housings, were locked under the press body pillars by a rear mounted air cylinder. The leaf was provided with a cast iron counterweight, and pivoted in bearing blocks extending from the rear of the press body. It was opened and closed by an air cylinder connected by a pivot arm keyed to the leaf shaft at one end, secured to the press stand at the other end by a pin and bracket.

Two hand operated air valves, secured on the top face on each of the two front upright press pillars, ensured that the press operators hands were clear of danger during the press closing sequence. Gentle operation of the valves controlled the leaf closing, and when almost resting on the front leaf stops, another valve was automatically actuated causing the bridge blocks to slide forward and engage under the four pillars.

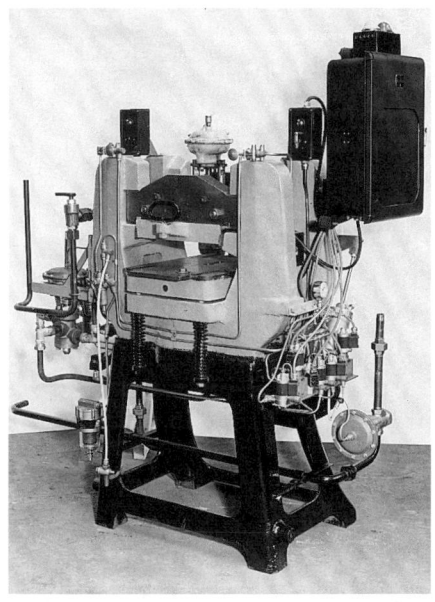

A fully renovated 7" press complete with Bristol Timer, remotely controlling the solenoid valves.

When fully forward the rear bridge block contacted an electrical switch, that initiated rotation of the Bristol timer camshaft. Micro-switches were actuated by cam shaft dogs to energise or de-energise six solenoid operated air valves, which on later presses were mounted on a common manifold.

The valves, in turn, controlled the following:

Valve 1	Steam, Diaphragm operated, Valve
Valve 2	Steam Trap, Diaphragm operated, Valve
Valve 3	Water, Diaphragm operated, Valve
Valve 4	Hydraulic, Diaphragm operated, Valve
Valve 5	Press Open Air Cylinder
Valve 6	Centre Pin Air Cylinders

The Bristol timer and solenoid valves were mounted on the same side of the press, the steam/water manifold assembly and hydraulic valve assembly on the opposite side. Presses were "handed" to suit installation requirements. A right hand press identified the location of the timer and the solenoid valves as being on the right hand side of the press body.

7" presses were used together with an air press equipped with a tool which punched the surplus flash from the pressing and, simultaneously, pierced the optional centre hole. The 12" press, however, was equipped with a hot knife trimming machine, mounted on the front upright press pillar, on the same side as the Bristol timer.

These presses required the following services for the moulding cycle:

140 psi	Steam
140 psi	Water at 25 degrees Centigrade
2000 psi	Hydraulic for 10"/12" diameter rams moulding 10"/12" LP records respectively.
or 1250 psi	Hydraulic Soluble Oil for 10" diameter ram moulding 7" standard & EP records.
100 psi	Compressed Air
240 V. 1 Ph	Electricity Supply

Mould block developments over the years had reduced press cycle times. They declined from 60 seconds for a 12" double press operation [two presses per operator], to 30 seconds during the final years of manual operation, when single press manning prevailed.

Two manual presses were installed in the development department in the early days, and were primarily used for mould block developments that eventually led to these very fast press cycle times being achieved. Bert Peasland was seconded from the pressroom and became involved in the 12" cycle time development programme.

At a time when 12" manual presses were operating on 40 second press cycles, Bert was successfully keeping up with his press, producing fully commercial non-classical product on a 24 second press cycle. This set the standard for the 12" 'Hayes' auto presses that were yet to be built. Similarly, Matt Harry was able to operate the 7" press on a 15 second cycle, including the trim and pierce operation! These fast press cycles were only made possible by the introduction of bi-metal mould blocks. These utilised beryllium copper for the cover plate, and stainless steel for the base block.

7" air press equipped with the standard pierce and blank tool. The removal of the lower panel for the photograph, exposes the control valves and the toggle action of the press.

This photograph of a left hand version of the 12" long ram press clearly shows the hot knife trimming machine mounted forward of the Bristol timer.

The major problem with these moulds was one of galvanic corrosion affecting the lands on the underside of the concentrically grooved beryllium copper cover plate. This was solved by electroless nickel plating a 0.002 ins. deposit on all exposed surfaces of the underside of the cover plate, but excluding the centre boss. Although slightly adapted from these originals, and with some further modifications, these moulds were adopted not only for use with the automated presses yet to come, but also as the Hayes standard. As such they were supplied to most overseas group factories.

The press operator's task was considerably eased by progressive developments concerning the methods used to pre-heat the P.V.C. When LP production was first introduced at Hayes, the biscuit of material was pre-heated on a steam heated platen, which was fitted with a hood equipped with infra-red heaters. Approximately six "biscuits" of plastic were on the platen at any one time, each with the appropriate pair of labels placed loosely on top. The heated biscuits were used in strict rotation and hand-rolled cigar style by the operator. They were then folded in half and placed in the centre of the bottom stamper on top of the label. The other label was affixed face upwards to the protruding top centre pin.

In the decade commencing in the late fifties, many new methods and processes were tested. The RCA Boomer was the first new-comer, and consisted of a vertical steam jacket chamber with a hydraulic ram that forced granulated material through a nozzle. The ram stroke was adjustable to vary the shot weight. Only a few of these machines of both 7" and 12" formats were used at Hayes.

The installation of granulating equipment in the weighing and grinding department, where the P.V.C. was compounded, led to the introduction of the "knife type" granule pre-heater in 1961. Of simple

To cater for ever increasing demands, a new line of 12" presses was installed in the mid-60s, and were used with single shot granule heaters.

construction, the units were provided with filtered inlet air fans connected to tubular electrical heaters, which passed thermostatically controlled hot air to a chamber of granulated material. This chamber was continuously fed from the hopper that was mounted directly above. The heating tube was closed on the underside by a knife-edged double platform that was out-stroked to allow the heated P.V.C. to contact the surface of the lower platform. The in-stroking action cut a slice of material of the required volume. It was this action which made these machines so unreliable and troublesome, the hot P.V.C. tending to stick to the knife, and thus distorting the shot.

The single shot granule heater prototype made its appearance in 1962. This consisted of a 15" diameter aluminium turntable sandwiched between upper and lower rectangular fixed plates. The turntable held four stainless steel headed bushes that produced pre-forms for 12" presses. Although intended principally for use with 12" presses, these machines could be adapted for 7" work. In its final form, it utilised the same hot air heating system of its predecessor, although heating efficiency was considerably improved by recirculating the air. Two separate heating stations allowed approximately 70 seconds total heating per pre-form. Each heating station was equipped top and bottom, with radially grooved P.T.F.E. discs concentrically grooved on their reverse sides to form air-ways through the discs. The hand filled hopper was located immediately after the shot delivery chute, which was at the front of the machine, and the turntable was manually rotated in an anti-clockwise direction. These granule heaters were simple, effective and reliable, and continued in use until displaced by the Werner & Pfleiderer [W & P] extruders in 1967.

Meanwhile, the U.S.A. manufactured Junior Egar extruder was purchased in limited quantities, and ultimately used with all press types except 12" automatic presses. An example of the larger version, designated Senior Egar, was acquired for comparative trials with the prototype W & P extruder in 1965. The outcome of these trials proved the Werner & Pfleiderer extruder to be vastly superior in compounding over a wide range of material

A single Junior Egar is used here to feed two 7" manual presses.

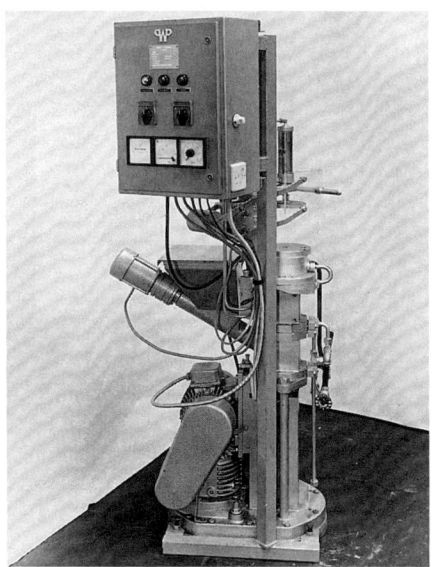

The Werner & Pfleiderer ZDS-P53 equipped for use with 12" manual presses.

formulations. The Senior Egar was returned to its manufacturer and the W & P, type ZDS-P53 was purchased in vast quantities, and became the standard equipment for 12" production on manual and automatic presses alike.

This was a significant development, for this extruder, with its twin screw configuration, fitted with kneading elements, and designed to extrude 1000 gms per minute, used dry blend formulations. The weighing and grinding department was ultimately totally transformed with automatic weighing, dosing, mixing, and air veying equipment, reducing manning levels to a fraction of those in the past. The need to produce both biscuits and granules had disappeared. No longer was the term "the black room", a legacy from the dirty nature of the manufacture of shellac stock, appropriate for the new weighing and grinding department at the Uxbridge Road site.

AUTOMATION OF THE 7" PRESS

The prototype 7" autopress in the new development department.

Around 1957, Hayes experienced its first sighting of an automatic record press, in the form of an in-line machine, which arrived from RCA as a result of the close collaboration that then existed between the two companies.

During 1960 a new Hayes designed 7" autopress prototype, based upon experience gained from the RCA machine, was installed in the recently established development department. It was equipped with the RCA Boomer to produce the shot of P.V.C., and utilised both the RCA press body and stand from the earlier prototype. A toggle air press held a tool that blanked the record to size and inserted the optional large centre hole. Twin steel belts, equipped with grippers, transferred the moulded pressing through these various in-line stations.

An initial batch of six machines to the same specification as the original were assembled in the development department, they were later converted to rack and pinion transfer mechanisms. A further 18 production machines followed during 1966 and all were equipped with the Junior Egar extruding machines, with their base stands cut off, to make it easier to mount them on the press frame.

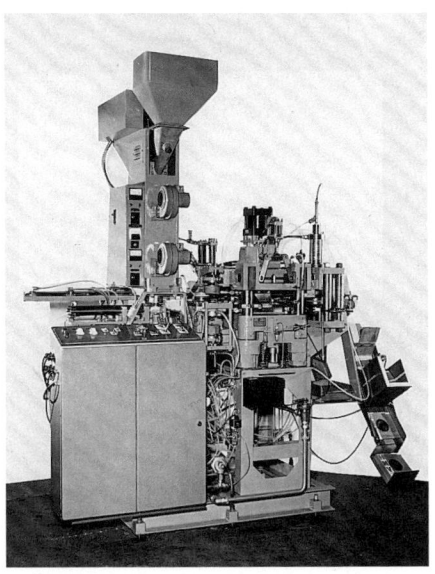

The final version of the very compact 7" Autopress.

A quantity of these machines, fitted with large centre hole tools to suit the continental market, were also manufactured for Pathe Marconi in France.

These presses, however, were destined to operate at Blyth Road only until production ceased there in the summer of 1972, and they were replaced at both the Uxbridge Road site and Pathe Marconi by the SP130 injection/compression presses.

THE SP130 AUTOMATIC 7" PRESS

In the late 1960s considerable interest was focused on injection moulding machines as the basis of a new generation of 7" automatic presses.

At Blyth Road, initial interest in the Metalmeccanica injection moulding machines gave way to the locally produced Windsor SP6 machine. A variant of this model, designated SP130, was eventually produced to EMI's requirements. It had a 130 ton lock, which also made it suitable for 12" moulding. EMI equipment was added to enable record production and various versions were manufactured, designated as follows:

Model 20007 - 7" injection/compression machine

Model 20012 - 12" injection/compression machine

The initial Windsor SP6 injection moulding machine in the Blyth Road development department.

Model 20712 - Dual function press converted from one mode to another in approximately 3 hours

When the new manufacturing facility at Uxbridge Road in Hayes opened in 1972, it was equipped with 26 of the 7" version. This total increased to 39 within a few years.

These presses were built by the machinery supply division, which was established in the toolroom. It employed, at its peak, approximately 50 skilled staff, consisting of machine tool fitters, millers, turners, grinders, and the like.

Many identical presses were constructed for Group overseas companies, including France, Italy, Sweden, India [Bombay], Eire and Mexico. Italy was also the recipient of several Model 20012 presses for 12" production, and Eire received the convertible Model 20712 for their fluctuating market demands.

RCA Records purchased thirty-one of the 7" versions, six for their U.K. factory at Washington, Co.Durham, with the balance for their main plant at Indianapolis, in the U.S.A., accompanied by a single 12" machine.

One of the SP130 – 7" versions installed at Uxbridge Road.

The 7" injection/compression moulding method was developed within two constraints. The new process had to be capable of being very simply introduced into a factory's operation. That is, it should not require a major re-engineering of the factory lay-out, or operation. It also had to be achieved with only minimal changes to established and traditional materials, i.e., stampers and labels.

The SP130 machines eventually designed had the additional advantage that they utilised P.V.C. material recovered from 12" production record trimmings. Cycle times were in the order of 15 seconds, although if well maintained and carefully tuned, presses could achieve 13 seconds. The process was noted for its low steam consumption.

Pathe Marconi developed their machines to operate as direct inject presses, utilising specially compounded P.V.C. materials of low viscosity. When acceptance was gained from the French record labels for the painted/ink label, the press cycle was greatly simplified by dispensing with the label loading and flash trimming processes. The moulding cycle was thus reduced to approximately 11 seconds, with the finished pressing being ejected into a chute when the moulds opened at the cycle end. An elevator loaded the pressing onto a record stand that, when full, was manually transferred to an independent inking station. This machine inked the label area of the record that had the information already moulded into it. The inking machine also incorporated a record bagging station.

The cost benefits of inking at the press were capitalised on when this system was later introduced at Hayes, but with self contained inking units fitted to each press.

Eight of the 7" SP130s originally installed at Hayes, and four of the Pathe machines [sent to Hayes when the French vinyl plant closed] were subsequently equipped with press opening robot actuators. The presses operated on the same direct inject principle, and this resulted in fast cycle times with individual press outputs of up to 2800 pressings per 8 hour shift. Another two of the Pathe machines were further converted to produce large hole discs, for juke box operators.

The aural standards of these pressings closely approached those of their 12" counterparts, but the presses had more demanding mechanical standards than their predecessors.

AUTOMATION OF THE 12" PRESS

Within the EMI Group, Capitol Records in America, had pioneered developments towards automation in the production of gramophone records. Their original automatic 12" presses were created around the existing manually operated presses that were used in the early 1960s. The same principle was adopted at Hayes, utilising a press mounted Senior Egar extruder bolted directly to the left hand side of the press, complete with "shot" [or plastic "biscuit"] transfer mechanism. The label magazines and record transfer mechanism were mounted on the other side.

The record take-off arm was pivoted immediately above the front right hand press shoulder. The tilt head press was arranged to open at the end of the moulding cycle, allowing entry of the record take-off head immediately above the pressed record, which was held on the bottom mould. A flash release actuator allowed the pressing to be lifted by the vacuum take-off head and transferred to the trimming station. Here the flash was removed by a wheel trimmer. After trimming, the arm was used to transfer the pressing on to the record stand situated midway on the arc between the bottom mould and the trimming station.

As the record take-off head removed the moulding, the label loading mechanism then in-stroked to position labels to both moulds.

The experimental nature of the prototype 12" Hayes Auto Press is apparent in this photograph with bundy and nylon tubing everywhere. Note also the truncated Senior Egar.

As it out-stroked from the press, the shot loading mechanism presented a shot of material centrally onto the label on the bottom mould. Once this was clear of the press, the tilt head closed and the new press cycle commenced.

It was soon realised that considerable cycle time reductions would occur by combining label and shot loading into one operation. By dispensing with the tilt head opening and arranging instead for the bottom mould [complete with record] to slide forward just clear of the top assembly, further time savings could be made. This principle was adopted for the production batch of 24 machines that were constructed around 1966/67. Fitted with Werner & Pfleiderer extruders, instead of the Senior Egar, they operated on 26 second press cycles and, surprisingly, lasted for nearly 10 years before being ousted by the new 1400 type presses.

They had proved reliable machines but suffered from the disadvantage of the pressing remaining in contact with the bottom mould for several seconds after leaving the top mould. This is not an ideal situation for precision moulding, and caused several quality problems.

The origins of the 12" Hayes Auto Press are clear in this photograph showing the machine installed at Blyth Road.

THE TYPE 1400 12" PRESS

The prototype 12" automatic press designed by EMI Electrola.

In the period 1966 to 1968 EMI's German factory in Cologne and Capitol Records in America had both developed fully automatic in-line 12" presses to suit their own requirements. The Electrola press from Germany was extremely sophisticated, being built around a double-acting four pillar press and equipped with the efficient Werner & Pfleiderer extruder. The Capitol 1300 type press, however, was very basic, built on their principle of KISS ["Keep it Simple, Stupid"!]. It utilised a four pillar up-stroking press.

In late September 1969 trials with both machines were conducted at Hayes, to determine which of them would best suit its needs. The over-simplification of the American machine, coupled with the limited efficiency of its single screw Egar Compounder, eventually told against it. On the other hand, the complicated control circuitry of the German press was considered, by management, to be beyond the abilities of the Hayes maintenance team.

The outcome of these trials was the creation of the 1400 press, which adopted the best features of both presses, but was more akin to the Electrola press. Like both press types it incorporated a sled, which, when out-stroked, simultaneously effected the following operations:

> Transferred a shot of extruded material complete with upper and lower labels from the extruding/label loading station to the moulding station.

> Transferred the previously moulded pressing, retained by having its flash moulded into the sled grippers, from the moulding station to the flash trimming station.

> Transferred the previously trimmed pressing, by means of a vacuum head, from the trimming station to the transit box.

Separator discs [initially circular but later square with chamfered corners] were carried above the sled. After every fifth pressing, a vertically mounted air cylinder, with a vacuum head positioned directly above the transit box, picked up a separator disc. It then out-stroked to position it in the transit box, once the sled returned to its normal position. This was to help the records remain flat during the critical post pressing cooling period.

The 1400 was to become an extremely reliable and efficient press, and over 150 were eventually built at Hayes, in the disc factory's own toolroom [known as machinery supply division]. One of the initial batch of six went to Toshiba-EMI plant at Gotemba, Japan in December 1971, and was subsequently used to manufacture, locally, twenty-one further machines [maintaining the Japanese reputation for copying everything!].

The Uxbridge Road factory was eventually equipped with 76 of the type 1400 presses by 1977/78. Early operating experience, however, identified certain areas that required design improvements. Maintenance problems occurred with the hydraulically operated centre pin cylinder, and this, coupled with the occasional hydraulic hose failure [oil everywhere!], lead to the substitution of a 5" diameter pneumatically operated cylinder.

The wheel-type flash trimmer was also a source of problems. At best, an acceptable edge finish was achieved but invariably it was below standard. Some improvement was attained by spring loading the cutting blades into contact with each other. The ultimate solution,

An early photograph of the 1400 press installed in Row 1 at Uxbridge Road. The press is already equipped with the pneumatically operated centre pin cylinder but the extruder retains the original and troublesome crosshead units. Note that 12" manual presses are still installed on the 'odd' side of the line.

however, proved to be the introduction of the time-proved hot-knife trimming unit to this press, a feature of manual presses that had earlier been abandoned.

The original main ram seals yielded poor performance compared with that achieved by their 7" SP130 injection moulding machine counterparts. Seal life expectancy was eventually greatly increased by the introduction of polyurethane 'U' seals, with some presses achieving over 2 years service before replacement was required.

On the 1400 press, the top centre pin performed the dual function of retaining the shot in position between the moulds, and forming the centre hole of the record. The long stroke of this pin resulted in heavy usage of them. This was largely overcome by increasing the amount of bearing for the pin in the body of the centre. This provided an adequate guide for the mandrel, which connected the pin to the actuating cylinder, and considerable increases in the life of the components resulted. Simultaneously, a cam arrangement, actuated by the rising ram, ensured that the air pressure on the out-stroked cylinder was relieved.

The selection of alternative materials for the centre pin body, particularly nitriding steel, which yields an extremely hard working surface, coupled with improved centre pin materials, justified the increase in the manufacturing costs of these parts.

The standard beryllium copper cover plate mould, originally used in latter years on the 12" manual presses, had continued in use on the 1400 press. Further design changes to the base block, made to overcome the problem of mould failure due to cracking, proved very successful.

The wide land profile of these moulds allowed the reintroduction of vacuum, to ensure intimate contact of the stamper to the mould face throughout the moulding cycle. This brought with it the major benefit of a reduction in the occurrence of "air marks". This is a fault where the record is incompletely moulded [i.e., there is air between the stampers, in places, rather than material].

A new design mould from Electrola eventually superseded the beryllium copper mould in 1982. This mould incorporated a brazed, spiral grooved, stainless steel cover plate, where 'O' ring life was considerably extended by the improved method of retaining the cover plate to the base-block. This was achieved by using a multitude of cap-head screws over the whole area, supplemented by the conventional centre boss retaining nut.

A post-pressing centring device was designed to cater for inaccuracies arising from the transfer of the pressing from the moulds to the turntable trimming station. Whereas previously an inaccurately transferred pressing resulted in a press stoppage, with this system the pressing was recentred by a long tapered turntable pin before being reclamped and trimmed. Pneumatically operated flash strippers were provided to reduce further the movement of the pressing on the turntable, as the sled grippers returned to the moulding station. By dispensing with the previously used fixed position flash strippers, and introducing pivoting ones, flash trimming problems were greatly reduced.

The 12" spiral grooved and brazed cover plate mould is renovated on the Capitol lapping machine.

This W&P extruder has the new improved crosshead unit ready for installation.

To enable the presses to operate effectively on either virgin powder or a combination of powder and reclaimed materials, a shot by volume device was provided. This ensured that the correct quantity of material was presented, regardless of the extruders delivery rate.

The cross-head units fitted to the Werner & Pfleiderer extruders were a constant source of trouble, as they suffered from continual mechanical failures. Long experience with this extruder without cross-head units, as used with manually operated presses, however, had proved the extruder to be generally reliable. In an effort to dispense with the troublesome cross-head unit a pilot batch of six 1400 presses were equipped with horizontally mounted extruders, which proved not entirely successful. Meanwhile EMI Electrola were tackling the problem by redesigning the cross-head unit with a more substantial unit, incorporating a vertical gear-box with overhead motor-drive. All the 1400 presses were eventually so equipped.

By the mid 1970s, a box shuffle unit was provided to allow presses to run on when the previous transit box had been filled, by replacing the full box with an empty one. With manning ratios then of one operator to four presses, this device allowed operators the freedom of completing the task in hand, i.e., stamper change, etc., in the knowledge that records would not be piling up and jamming on other machines in their charge.

A trial batch of presses were fitted with bagging equipment, which inserted the pressing into its bag at the press, thereby reducing manufacturing costs. It also avoided damage to the product that invariably occurred when un-bagged records undertook their journey from the pressroom to the examination department, by means of an overhead conveyor and a high level roller conveyor.

This proved highly successful and eventually all 59 presses that could be fitted in the revised pressroom lay-out, necessitated by the addition of the bagging unit, were so equipped.

One further major design change occurred as a result of Hayes' involvement with the 7" press, which utilised the Windsor SP130 injection moulding machine. On this machine, the main ram piston was out-stroked by a relatively small diameter inner piston. Large poppet valves were opened during this movement to allow oil to be drawn into the main ram cavity. At the end of the stroke, the poppet valves closed and the trapped oil was then pressurised to the required level. This principle was applied to the 1400 press and fortunately coincided with a press bottom casting replacement programme, brought about by failure of the originals due to cracking. The new casting was provided with the additional oil way. A special valve to fulfil the function of the poppet valves was designed and manufactured for EMI by Hale Hamilton Ltd., specialist hydraulic equipment manufacturers, at Uxbridge.

The clear perspex rear guard panel was removed for this photograph of a "hydraulic saving" equipped 1400 press.

With the addition of an oil reservoir rigidly connected to the Hamilton Valve, and the replacement of the 1" diameter hydraulic valve by two one-quarter inch diameter manifold mounted valves, the modification was complete. This reduced the use of pressurised hydraulic oil from approximately 1 gallon per press cycle, to 0.28 gallons per cycle, resulting in major energy savings in subsequent years.

The 1400 press continued in use throughout most EMI plants and, with further minor modifications and improvements, could truly be described as the "workhorse" of the Company, seeing it into the 1990s.

MUSIC TESTING EQUIPMENT

The testing of records, traditionally carried out by operators listening to them either in small studios on loudspeakers, or on headphones, was identified by EMI management as a very subjective and expensive way of undertaking the task. Despite this, however, it

The music test department equipped with the Mark III disc defect detectors.

was not until 1965 that an automatic testing machine was developed by Mike Batchelor, at the Central Research Laboratories in Hayes.

The machine initially developed tested an album at real time [33 1/3 rpm], giving a read out of defects. In a later development, the electronics were improved to allow the record to be played at 78 rpm, thus reducing the time taken for the test. Both machines played the record in reverse, from the centre out. These units were named Disc Defect Detectors, and existed as mark III [the 33 1/3 rpm machine], and the mark V [the 78 rpm version]. They were, however, affectionately known throughout the factory as DDDs.

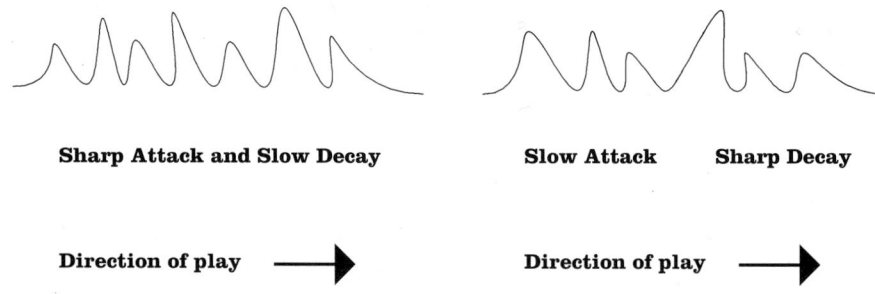

Fig. 1. Waveform of Music.   *Fig. 2. Waveform of a typical defect.*

The electronics within the machines analysed the wave form on the record. The wave form of music normally has a sharp attack, and slow decay [see fig. 1]. Most defects on records, known by colloquialisms such as snap, crackle and pop, have wave forms having a slow attack and sharp decay. Thus when the record is rotated backwards the sharp decay of the defect can be electronically detected [see fig. 2].

There were three other techniques used for testing vinyl product, all of which were carried out in the time honoured manner. Classical records were fully tested [played all the way through], semi-classical were closely spot checked, and pop were spot checked, including the run-ins and run-outs. Testers distinguished classical/semi-classical records from their matrix catalogue number, for example the prefix **ASD** indicated a classical product and **EMX** a semi-classical.

Besides these checks, records were also tested for eccentricity, on a machine devised by Ray Saunders.

Classical records had always been made, in Uxbridge Road, on specially designated presses. The position in the pressroom of these, however, has changed over the years, as have the particular presses concerned! In 1983 the designated presses used specifically for classical pressings were changed once again. The location altered from the bottom right hand corner of the pressroom, to the first five 1400 presses of the even side of row three.

To overcome the problems then being experienced with the reject rate of classical pressings, running at over 20%, these presses were subject to special attention from the quality control department. A process of tackling some of the problems began. The lighting was enhanced over these presses to help identify visible faults, and there was a special white painted "egg crate" ceiling installed. The effect of this is uncertain, but it certainly made that part of the pressroom look "upmarket"!

As orders for classical records were so small, from 50 to 500 per title, a communication system based on indicator lights was introduced between the pressroom and music test. This was designed to ensure that the records were tested before the job was finished! The indicator light system was to be used for all new jobs, stamper changes, and all finished jobs, the points at which testing had to take place. When any of these events occurred, the light system, which comprised three lights for each of the five presses, was used thus:

      RED LIGHT   Records require testing

      YELLOW LIGHT   Records being tested

      GREEN LIGHT   Records passed inspection

The pressroom only used the red panel button, which then lit up on the music test light panel to show that records on the press required testing. The music testers, who were responsible for testing the classical product, would respond immediately.

Once the records were collected from the press, the red light would be cancelled and the yellow light would be lit, and only after passing the test would the green light be activated.

The tester would listen to all records with the use of headphones and used a gauge that was sub-divided into 30 sections covering the playing area of the record. Each gauge division would be "spot-checked", skipping through loud passages and dwelling on quiet ones, listening for any faults they might pick up. Each new tester was given both a hearing test and full training on what faults/defects to listen for! On a semi-classical record, only some of the 30 divisions of the gauge would be checked, and on a pop record fewer still.

If a defect was found, testers would mark it off and refer it to a supervisor/senior tester, once they had completed the entire test on

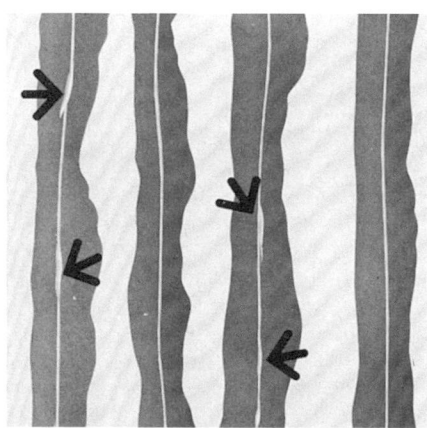

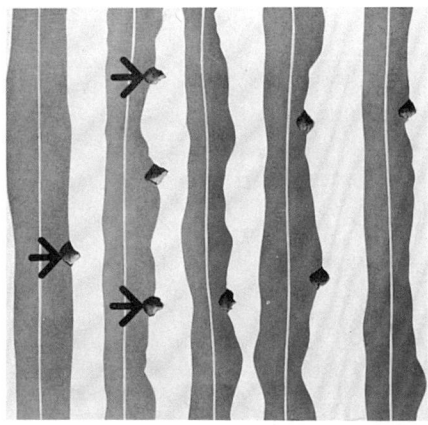

Fig. 3. **Apex-Damage**

Also referred to as 'stamper-touching'. Damage to the apex of the groove normally at Gauge 2 to Gauge 8, occasioned by the two stampers touching before the plastic material fills the mould cavity. Occasionally seen as a concentric pattern by reflected light, but not always visible without magnification.

Fig. 4. **Non-Fills**

Also referred to as 'airmarks'. A cavity on the outside groove wall, occasioned by incompleted moulding.

May be seen as a random pattern or as radial streaks.

Will vary from pressing to pressing.

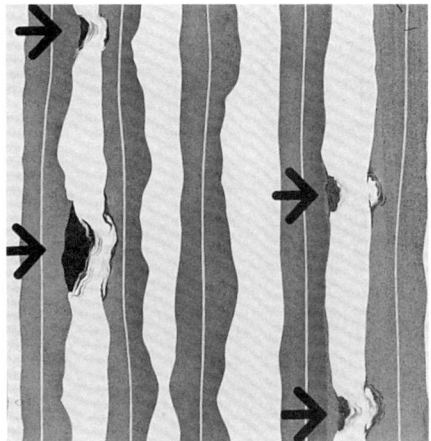 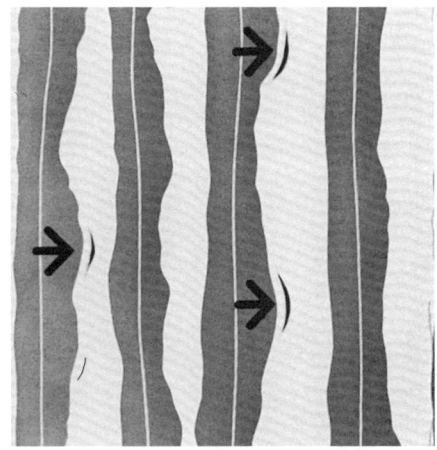

Fig. 5. **Pre-Release Damage**

Also referred to as 'stitching'. Localised damage of the outer groove wall, occasioned by shrinkage of the disc before the stamper has completely released from the plastic.

Seen as a concentric pattern by reflected light.

Fig. 6. **Post-Release Damage**

Also referred to as 'stitching'. Localised damage of the groove shoulder and land, occasioned by the stamper coming back into contact with the disc immediately after release.

Seen as a concentric pattern by reflected light.

that disc. The record would then be played on higher quality audio equipment and the defects analysed. Occasionally the use of a microscope was necessary.

Typical examples of defects are shown in figures 3,4,5,6.

If a defect was found on a classical or semi-classical record the process was slightly different. The 6 decks used by the music test for testing classical/semi-classical records, had an inter-com system attached. This enabled faults or queries to be communicated to the senior tester, with a verbal description, as soon as they were detected. Once the senior tester had finished the analysis they would then feed back the results to the tester.

With the decline in demand for classical vinyl records that occurred with the introduction of the compact disc, it was decided to manufacture all the European requirements in one factory. Classical record production at Hayes was gradually transferred to the Cologne factory, and the last remaining titles went in 1989. The loss of this production was, however, more than offset by the growth in the new 12" 45 rpm disc. Uxbridge Road benefitted from having the former classical presses to help cope with them.

Headphones were introduced into the music test department in 1962, the first to be used were made by AKG, model no. K150. These were updated in the mid 1970s with the AKG model no. K160.

In 1982 the matrix quality controllers changed over to smaller headphones made by Pioneer, model no. SEL 5 [similar in appearance to Walkman headphones]. The pressroom quality controllers, however, tried them but preferred to keep the AKG K160s. In 1989, Sony produced their type MDR 32 headphones [similar to the Pioneer SEL 5]. These headphones were tried and tested by the pressroom quality control, who decided, finally, to change from the old K160s to the MDR 32 headphones, as they were smaller, lighter, and produced a more accurate sound.

In 1989, to enable EMI at Hayes to comply with new Health and Safety laws being introduced on the 1st January 1990, Peter Dix of the Central Research Laboratory was assigned to monitor, test, assess and adjust all of the listening equipment. This was crucial, as the new laws affected the way in which records were tested. The volumes at which records were listened to were closely controlled, especially the 7" and the 12" 45 rpm discs, as these were generally recorded at a high level. In order to comply with these laws, the level at which 7" discs were checked was reduced by 7db.

Responsibility for visually checking the records remained with the supervisors in the pressroom until 1987. Then the structure of the music test department changed, to include visual [roving] inspectors. Their duties were to monitor, continually, the quality of the product as it was made, involving visits to each press. With the arrival of T.Q.M., however, and changes under the Framework for the Future programme, press operators became responsible for the visual inspection of product from their presses.

Vinyl records were monitored not only for audio quality. Part of the Framework for the Future programme, in 1989, saw the introduction of added checks. These involved record weight, and, for 7" product, visually checking the edge and inking quality. The monitoring results were used for analysis and subsequently for corrective action and long term improvement to the processes and systems.

The introduction of D.M.M. processing and the purchase of the licence to use the D.M.M. logo entailed strict quality control procedures, and the rumble and surface noise were then measured to agreed standards, using the RUMS 77 rumble meter.

After the major reorganisation undertaken in 1992, when the workforce in the vinyl plant was reduced from over 200 to 36, the quality control department disappeared. Press operators became responsible for undertaking testing procedures for the product from the presses in their charge.

PACKING EQUIPMENT

The record examination department was, in the late 1950s, extremely labour intensive, employing approximately 140 to 150 female staff on both morning and afternoon shifts. An evening shift also operated for approximately nine months of the year, coinciding with peak production demands. One hundred and ten tables were provided, alongside the desk height Flowlink Carousel conveyor system, which delivered the jobs to the operatives for routine 7" and 12" product packing. The finished and boxed product was then placed on an overhead belt conveyor and, having passed the cursory checkers inspection, continued its way to the record stores. Additional staff were utilised on box making, box set make-up and resleeving operations.

Examination and packing of 12" records was a labour intensive operation. Identified carrying out this task are (left to right) May Taylor, Evelyn Matthews, Kit Brotheroe, Lily Borwick, Agnes Shaw, Mary Bass.

All 7" product was received from the pressroom on record stands, each containing 75 records, ready to be inserted by hand into their bags. An urgent task for the development department was to automate this process, and a simple, but very effective, machine was constructed for this task. Fourteen such machines were ultimately built but they proved so effective that normal demands were accommodated on six machines only. Output levels of 4,000 per hour were achieved from a team of 3 operators using 2 machines.

Machine operation was very easy. The loaded record stand was located on a spring-loaded platform that centralised the stand pin with that of the double hinged machine pin. With the pins aligned, the operator lifted the records on to the machine pin and pivoted it to its operating position slightly downwards from the horizontal. The records naturally slid towards the end of the pin, but were retained by the autochange-style pin-end device.

With bags loaded in the magazine immediately below the records, a motor-driven rotating eccentric device caused the combined record and bag suction devices to pick up when out-stroked. Then, separate cam operated valves released first the record, and then the bag. The bagged record dropped via a chute into an open 25 way transit box. Built-in counters caused the machine to hesitate after 25 records had been bagged. This allowed the operator to slide a replacement box in position, and to reset the machine to continue bagging.

Following closely behind was development of a machine to semi-automate the packaging of 12" records. As it was customary and necessary to inspect visually each 12" disc, the design of the equipment had to meet this requirement. The machine operator inspected the records while removing them from the record stand, which held 55 records, with separator discs between every 5 records.

The quantity of 55 records per stand was selected as it catered for an anticipated reject level of exactly 10% leaving, in theory, sufficient good product for two complete transit boxes of 25 each. [This ratio of 55 records per stand was still in use until the stand and its successors ceased to be used in 1992. Even then there was still no apparent sense in its choice, as rejects never occurred at the rate of 5 per box; it was usually all or nothing!]

The operator placed the inspected record into a vertical slot, and allowed it to drop directly into an open poly-lined bag.

The bagged record was then released on to a chute that guided it into a horizontal plane between a pair of open sheet metal plates. The top plate closed to clamp the bagged record, and the assembly transversed sideways to enter the mouth aperture of the jacket. When fully entered, a vertically mounted air cylinder clamped the record, bag

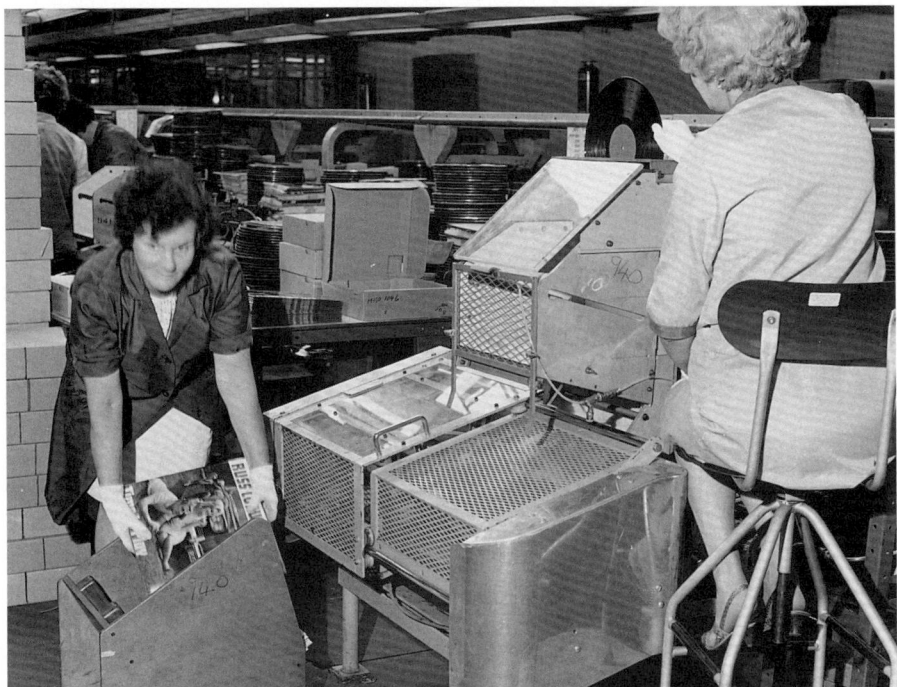

Russ Conway's latest album is packed by Maude Lee and Evelyn Matthews (seated).

and jacket as the sheet metal plates were withdrawn. The record assembly was then powered forward by rubber rollers as a replacement jacket was simultaneously fed from the jacket magazine. A freestanding magazine that received the finished product was periodically emptied by the second machine operator to be counted and boxed.

The jacket magazine was reloaded with multiples of twenty-five jackets from the next transit box, which could be done with the machine running. Inner bag reloading, however, was carried out with the machine stationary. Approximately twenty of these machines were eventually manufactured and installed in the record examination department, processing product at a rate of 500 per machine hour. This compared to an average of 100 per hour for an operator inspecting and manually bagging the product.

At the time of the introduction of these machines, 12" records were only inspected and bagged in the examination department and were then finished in the record stores, when jackets were received from the printers. These machines therefore allowed the product to be processed very much more efficiently than before. A simpler version of this machine, to bag 12" records only, and not insert them in the outer jacket, was also manufactured. Six of these were utilised for product that formed part of box sets, or required special packaging.

OTHER EQUIPMENT

An early arrival at the development department was the Hayes version of the Capitol record scrubber. This was a machine designed to remove completely the record labels from deleted or scrap 12" product, so that the material could be re-used. The records were soaked overnight in cold water before being fed into the machine, and passed through its rotating wire brushes.

In spite of this severe action, some label paper fibres remained ingrained in the record and the occasional strand of wire from the brushes did little to enhance material quality! After a couple of years use in the record returns area, the scrubber fell into disuse. The process returned to the previous practice of punching out the label area, and re-using the rest of the material.

The 12" de-sleeving machine, also from Capitol Records would suffer a similar fate. This equipment was designed to eject the record from its sleeve and jacket, and was vital in the American market where dealers returns of unsold product were uncontrolled. Again, it was used for a brief period in production but was not overly successful. This was possibly due to the higher quality of the packing used in the U.K., compared to that used in America. The machine was eventually consigned to the scrap heap.

The use of several Junior Egar compounding machines necessitated the investigation of improved means of reloading their hoppers, which were approximately 9 feet above floor level, and certainly too high to be hand loaded! The solution provided by the development department consisted of several Venturi hopper loaders. These were constructed in the works engineers repair shop, utilising a half-inch compressed air supply of 100 psi, to blow the material to the top of the hopper. They performed their task of conveying granules somewhat noisily!

When the 7" automatic presses were installed in the pressroom, they were fed with material from a blowing system installed in the weighing and grinding department. Based on systems used by Capitol Records in America, a continuous feed pipe inter-connected each press hopper, and deflection plates caused the granular material to enter the hopper. This was a continuously operating system that added considerably to the already noisy environment of the pressroom, but was a further labour saving scheme.

CASSETTE EQUIPMENT

In 1899 a Danish engineer, Valdemar Poulson, first had the idea of recording sound waves on steel tape. He called his system the Telegraphone, as it was intended to be used to record telephone messages. As the steel tape kept breaking, the invention was largely ignored until 1932 when Poulson, and a group of German engineers, found a way of recording electrical impulses as magnetic patterns on a coated plastic tape.

Just before the Second World War they improved the quality of their recordings by using a bias signal. This development produced better recordings, which were used for propaganda purposes at that time, although during the war steel wire continued to be the predominant medium utilised, as it was more durable in the field.

After the war, developments turned towards commercial tape machines for home use. These machines used plastic based tape, early experiments using paper having been abandoned. EMI was very interested in this development, and proceeded to plan the introduction of pre-recorded tapes for home use.

As a result of the predominance of vinyl records, and the need to relate this new medium to both the dealers and public, the name "tape records" was coined. It was in was 1955 when EMI introduced its first stereo 7.5 inches per second [ips] reel to reel tape records. They were issued on plastic spools, and were 7" in diameter, utilising 1/4" tape.

The tapes were recorded using BTR1 tape recorders. One was used for replaying the master, while the second machine copied from it. This mini production line was set up on the third floor of the cabinet factory in Blyth Road, Hayes.

Fred Brooks joined EMI in 1959 and was responsible for starting developments in high speed duplication of tape. Initially using BTR2 machines the increased ratio of 2:1 was achieved [i.e., the tape was recorded at twice its normal playing speed]. The retail price of tape records in 1959 ranged from £2.2s 0d to £4. 4s 0d each, and two versions were available, single channel and stereosonic. Another version introduced in 1962 was twin track mono pre-recorded at 3.75 ips, by this time, all tapes were being recorded at a 4:1 duplication ratio.

The various versions of the product were so confusing that the consumer needed to become a technical expert just to understand the different speeds, track layouts and lacing required, merely to ensure the music played in the correct direction! The time was right for an easily loaded tape format that was not so unwieldy in use.

In June, 1966 EMI released its first pre-recorded cassette tape record [later to be called musicassette]. This new format was easy to use, and the first production copies were manufactured by Philips, and checked in Hayes before release. The housings [known as C-Zeros] were white with a paper label, and initially the cassettes were available only in mono. The playback speed was 1 7/8 ips.

The early growth in musicassette sales encouraged the company sufficiently for it to make an investment in its own production equipment in 1968. Gauss, an American supplier, was selected for the purchase of the first duplication system. This comprised ten duplication machines, type 1200, with one loop bin.

Mike Lloyd, an engineer in the EMI cassette factory, was mainly responsible for the physical installation of the equipment. The 1" production masters were recorded in the studio, located on the third floor of the cabinet factory, by John Cox. His pipe smoking left one with the impression of leaving his studio, where he would describe the "black art" of transfer engineering, under a cloud of mystery!

The duplication ratio at this time had increased to 32:1. An Electrosound QC machine was used for checking the recorded tapes. At this time all tape was received on plastic spools and wound back on to them after recording.

No form of cue tone generators were then in use, so a section of pre-recorded cue tape was spliced into the loop master before production could be run. This was to ensure that the cutting/winding machines could find where one programme ended and another began. Electrosound flat decks were used to wind the individual cassette lengths into the C-Zero [a cassette housing with zero tape].

On completion of the wind, the Electrosound tape splicer was used to splice the ends together. This machine cut the splice width-wise from the splicing tape, leaving waste material between punched out sections and on both edges. Initially cassette labels were applied by hand, and then, again by hand, the cassettes were inserted into the library box, along with the inlay card.

In 1969 the first pre-recorded 8-track cartridges were launched. These were seen primarily as a means of sound reproduction for use in cars. The early success in North America of this new format prompted the launch. The cartridges utilised lubricated tape, all of which had to be imported from the U.S.A. The duplication ratio was 16:1, and the duplication process used the same equipment as musicassettes, but configured for 8-track use.

Further equipment was purchased similar to that for musicassettes so as to provide two independent production lines, albeit a smaller one for the 8-track.

The Board of EMI then took the decision to manufacture their own blank tape; thus EMITAPE was formed in 1969, and equipment installed to start production. This factory was located in Dawley Road.

As quantities increased from the initial hundreds per title, other types of production equipment were installed. The hand application of cassette labels ceased when a machine called the Pony Label Dri was commissioned. This machine used heat sensitive labels that were drawn across a hot plate during transfer. At this time it relied upon the operator manually turning over the cassettes for sides 1 and 2.

A new version of the record head came into use, using focused gap techniques. These heads were manufactured from ferrite material, and the base was optically ground in reference to the gap, to ensure the correct azimuth. The design of record heads has changed very little since.

King 600 series winding machines were introduced in 1969; these were heavy and stood upright on a work station in use. The C-Zero was laced into position by hand, and on depressing the start button, the leader tape was cut and the recorded tape was spliced and loaded into the cassette housing. The machine was electro-mechanical, and it used a cam shaft to provide the automatic functions.

Tapematic cassette winding machines type 2002. These machines replaced the King loaders. You can see the duplication area beyond the winders.

As with all new machines they were not without their problems; any engineers who have worked on the King 600 series machines will remember the "P valves". The replacement of these required the engineer to have tiny double jointed fingers, that could also act, simultaneously, as tweezers!

By the time the tape factory moved to Uxbridge Road, in 1972, further Gauss machines had been purchased, together with flat deck machines for 8-track production.

The initial impact on entering the tape records department at Uxbridge Road was one of a factory with rows of King machines set out in pairs, on work benches either side of a central conveyor. These delivered the wound boxes of cassettes to the Pony label machine at the end of the line.

Jim Wilsher [now Engineering Manager] joined in 1975 and remembers Min Emo using the Pony machine at great speed. The first machine had, by then, been modified by the manufacturer for automatic turnover, and label indexing magazines had been fitted so that both labels could be applied simultaneously. Min, however, seemed able to get it to go faster than the manufacturer's specified rate!

The Gauss duplication area ran in an L shape, with a further duplicating area in studio 4. This room housed a loop bin and ten

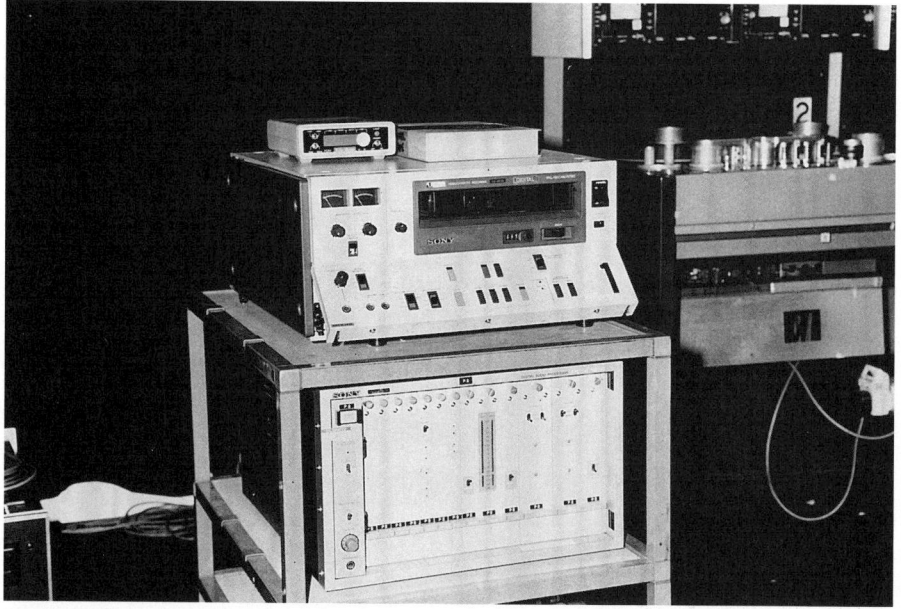

Sony Digital Umatic machine type VO 5630. These machines replaced the 1/4" tape machines that were used to replay the studio masters.

duplicating machines. By this time the blank cassette tape was being received on 1/4" Nab hubs, and the recorded tape was wound back directly onto the same type of hub. Small jockey wheels were used, in side guides, to help the winding, and these were pushed along the guides as the tape pancake diameter increased. The manager of the department at this time was John Simmons, later to be the director in charge of Uxbridge Road manufacturing.

Other areas of development in 1975 included the use of Dolby B encoding, for tape noise reduction, and the use of a new tape from EMITAPE called X1000.

An important investment made at this time was an automatic packing machine, called the "Ponzoni Vico", from Italy. This heavy machine consisted of a large turntable with four stations. Its mechanical cam system for packing cassettes, together with inlays, into a library box, was driven by a large motor and gearbox. A promotional video for this machine showed it in operation in the Ponzoni factory car park in Italy. The sun was shining and the machine appeared to be running very quickly. There are those among us who, having experienced this machine, think they must have speeded up the film!

In production the Ponzoni packer was to prove a headache. One item that gave trouble was nicknamed the "cat's whisker". It was a piece of thick piano wire, which helped the insertion of the cassette into the library box. The cry from the operator that "the cat's whisker has gone" meant that one side of it had broken off from the centre. This in turn would cause an engineer to appear with a large hammer and punch to fit another one - a task that was performed several times an hour!

The Ponzoni Vico machine was finally taken out of service in 1980, when the first Ilsemann KZM3 packer was installed. The Ponzoni machine was stripped down to its component parts ready to be rebuilt and assembled like new. This never happened, possibly due to the dislike the engineers had for its design, and the machine was eventually "sold as seen" in cardboard boxes, for more money than the original purchase price! The Ilsemann, however, is still in use!

The production of 8-track cartridges was beginning to be reduced as the cassette gained ground, not only for in-car use but also in other portable battery powered machines. As the quantities of 8-track cartridge requirements had reduced, another Pony Label machine was modified for cassette use.

The engineer from the Pony label company arrived to undertake this modification, installed the automatic turnover station, and asked for some test cassettes. Jim Wilsher stood by looking on as the controls were adjusted for correct operation. Many cassettes were being damaged in the process, and Jim became concerned, so he asked if the engineer could set the machine up without using cassettes, because of their value. The engineer replied that the cost was nothing, considering he had just installed a label machine in a Scotch whisky bottling works, and after twelve broken bottles he had been overcome by the fumes!

Eddie Wilcox was appointed manager in 1977, and during this time he created a research team headed by Fred Brooks. Paul Treanor joined shortly afterwards, and one of the finest developments made was the new winding arms for the Gauss slaves, in use on all 1200 series Gauss duplicating machines today.

In 1979 a major new installation took place, in the form of the Apex printing machine. These machines printed the title information directly on to the cassette without using labels. The Apex was originally equipped with a long drying oven to cure the ink, that was expensive both to purchase, and in power consumption. With the changes that occurred in ink technology, these eventually became unnecessary. Suddenly, the Pony label machines stood silent, as cassettes became directly on-body printed, using blue ink.

At this time there was a rapid succession of managers. Godfrey Burton replaced Eddie Wilcox, in 1980 and was, in the same year, followed by Peter Hall. His departure, in 1981, signalled the beginning of the reign of Geoffrey Grimmel, who joined as an engineer in 1975, which was to last until 1991.

Development continued apace, however, despite these changes. The 1980-81 period saw the introduction not only of the previously mentioned Ilsemann KZM3 packer, but also of 64:1 duplication ratios and the launch of classical musicassettes using the new BASF chrome tape. At this time 8-track cartridge production had completely ceased, and the equipment that could be modified for cassette production had been changed.

In 1982, an investment in King 790 winding machines was made, to replace the old 600 series. These machines were micro-processor controlled automatic cassette winding machines, and had no more 'P' valves, but used rows of low voltage valves instead. This was the first type of completely automatic winding machine made available for production use.

The following years continued the pattern of growth in terms of production quantities as well quality. EMITAPE was replaced by Capitol tape, manufactured by them in the United States. The use of C90 [thin base tape] meant that only two types of tape were in use for production, chrome and ferric, instead of the original mixture of C60 and C90, in both types.

XDR [Extended Dynamic Range] musicassettes were introduced in 1982, and were the first production musicassettes to use digital original masters for production. Other improvements made within the XDR programme were precision tape guides, and signal processing amplifiers.

During the mid to late 1980s Tapematic winders were introduced. These superior cassette winding machines allowed both the operator, and the engineer, to monitor the current performance and operation of each machine. One packing line became equipped with an Ilsemann CM1 overwrapper, which individually wrapped each cassette in clear film.

A permanent night shift was introduced in 1988, and in one month in 1988 over 3 million musicassettes were manufactured. By 1990 the King 790 machines had been replaced by further Tapematic winders, and the Apex machines no longer used the drying ovens, as air drying of the ink became available. The colour of the ink had changed to white, to provide more easily readable printing, on both clear and black cassette housings.

Over the years, improvements were made to the production floor layout, by making the whole area open plan, and investing in further production equipment, to cope with the increased volumes. In 1989 the decision had been made to change the open plan layout into a flow system, and expand beyond the brick wall that divided cassettes from the disc factory. This need for expansion was brought about because the musicassette was enjoying a firm market position with high volume demand, outselling both vinyl and CD formats.

These changes formed part of the Framework for the Future programme, which incorporated many ideas and plans to improve production service levels. The plan included the introduction of T.Q.M., coupled with the implementation of a programme to meet the required standards for BS 5750 accreditation.

From its humble beginnings the musicassette had now reached the centre stage, and was firmly under the spotlight.

In 1990 the first Ilsemann KZM4 packing machine was purchased, with a Sollas polythene bulk wrap machine, to increase packing capacity.

During 1991 further Tapematic winders were purchased, and the duplication ratio on the Gauss 1200 series machines [yes, the original ones!] was increased to 80:1, by using Electrosound loop bins.

In 1992 further improvements continue to be made within the now much larger, open plan, manufacturing area, yielding considerable gains in both efficiency and customer service.

THE POWER HOUSE

The power house could be considered to be as important to the plant, as the heart is to the body. It occupied the south east corner of the main building at the Uxbridge Road site. This part of the site was rebuilt with a high roof, to meet the needs of both the power house and the combined material mixing and scrap recovery areas. At Blyth Road all the power services were provided by the company's works engineering department, and all pipe-work and electrical supplies were installed and maintained by this central service function. Twenty five personnel transferred to the new site at Uxbridge Road from the works engineers department, including electricians, pipe fitters and mechanical trades.

The transfer of both disc and tape manufacturing, and distribution to the new Uxbridge Road site demanded the installation on the new site of a significant quantity of power services. These were under the control of the engineering manager. A Power House Superintendent, and four shift engineers were recruited, and from 1972 until mid-1992 a continental style shift pattern was operated, providing cover for 24 hours per day, all the year round. The shift engineers worked a pattern of 2 days and 2 nights on, followed by 4 days off. This system of 12 hour shifts yielded a 42 hour working week, and continued despite changes in shift patterns on the production side.

A large variety of equipment was operated in the department, and listed below are some of the major plant items under their control.

The chimney base is prepared for the already installed boilers in the Power House.

The steam boilers, in equipment terms, are the real heart of the plant, and at Uxbridge Road there were three boilers, with a total steam capacity of 66,000 lbs per hour. Manufactured by George Clark & N.E.M. Ltd., of Hartlepool, these Maxecon boilers each weighed 35 tons and operated at a pressure of 200 lbs per square inch [psi], fired on natural gas. A pressure reducing valve in the system allowed the pressroom supply to be maintained at 150 psi, while low pressure requirements of 40 psi were supplied via a further pressure reducing valve. Although principally supplying steam for production processes, they also supplied both the factory and the distribution centre with heating, including some domestic hot water for the distribution building and the works canteen.

Two boilers would normally cater for these requirements, but peak demands have at times dictated the use of all three. Flue economizers were fitted in 1985, which resulted in efficiency improvements of approximately 4%-5%.

The water treatment plant was installed to protect the boilers and pipe-work from corrosion and scaling, the plant ensures that incoming water supplies are suitably treated. Two plants are available and each can process water at the rate of 7,500 gallons per hour. They are not continuously available, however, as the process of regenerating takes two hours under normal conditions. The process involves lowering the water's alkalinity levels and then its hardness, using "ion" exchange resins. Further treatment, involving the use of four additional chemicals, tailor the water to suit the factory's requirements. The boiler house consumes up to 100 tons of chemicals per annum for this treatment.

The recirculated cooling water system is needed to cool down a variety of machinery, but is primarily used to cool down the moulds on the presses during the moulding process. The cooling tower tanks, which have a capacity of 52,000 gallons, are equipped with eight 6ft-diameter quadruple-bladed fans. These are designed to maintain the cooling water at 25 degrees Centigrade. The hot water returning from the presses is broken down into droplets by being poured into the top of the tower, over a "mesh", and then the fans move air through the droplets to cool them. Two high pressure pumps are used to supply the pressroom with cooling water at 150 psi, and three low pressure pumps supply other process requirements with water at 40 psi. Other pumps recirculate cooling water over the cooling towers to maintain operating temperatures, and yet more constantly discharge the underfloor pressroom common return tank. A total of eighteen water pumps are installed in the pump house, located directly under the cooling tower and tanks.

The recirculating cooling water system, used both for manufacturing and other purposes, can lead to losses approaching 5,000 gallons of water an hour on very hot summer days, through evaporation from the cooling towers!

The central hydraulic oil system is not used throughout the pressroom, as the Windsor SP130 injection moulding machines, used for 7" disc production, are equipped with their own electric motor driven hydraulic oil pumps. The 12" presses, however, rely upon a central hydraulic system for this service. Power presses used in the matrix and label printing departments also require hydraulic oil pressure.

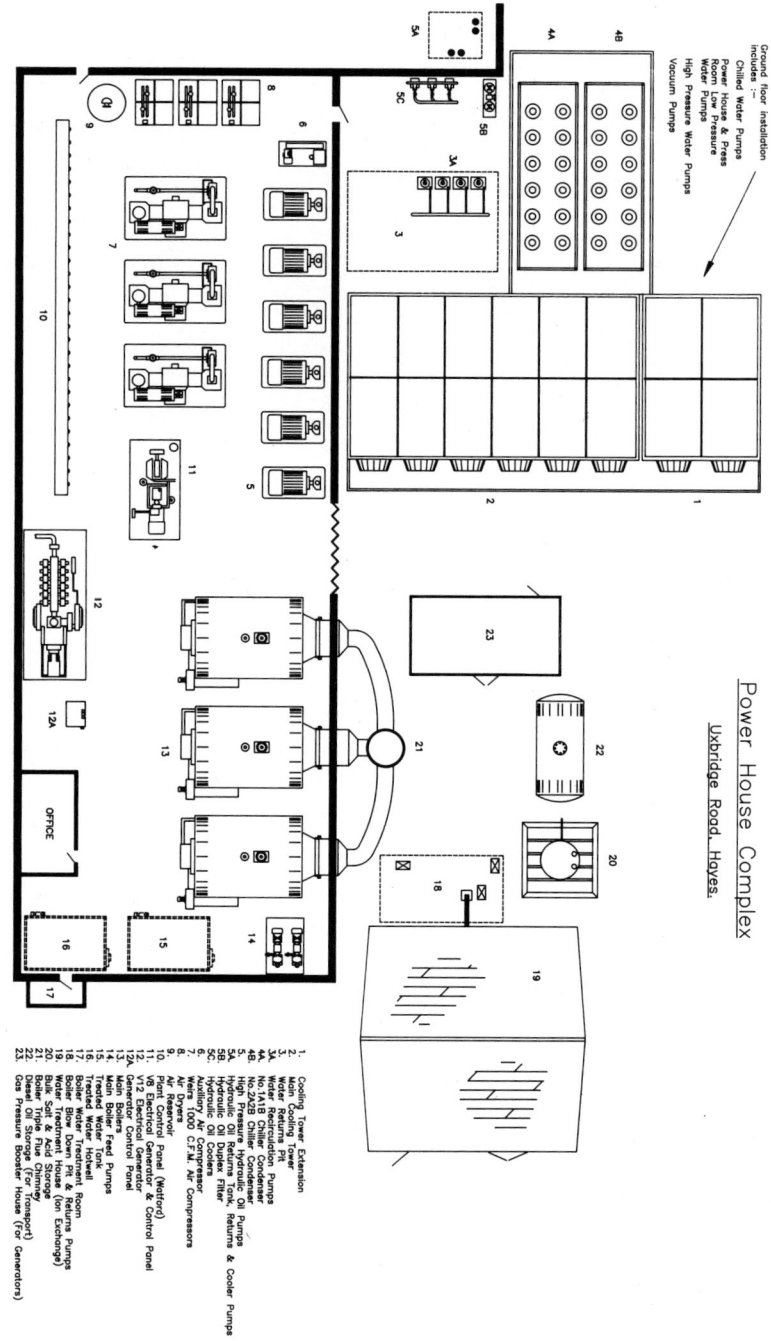

The central system is equipped with reserve tanks holding 3,500 gallons of oil and within inter-connecting pipe-work a further 1,000 gallons, under a pressure of between 1950 psi [minimum] and 2300 psi [maximum]. Twenty seven Greer Mercier bottle accumulators are located off the main feed pipe across the pressroom, each with a capacity of eleven gallons.

Six hydraulic pumps, each with a capacity of 40 gallons per minute, can be "put on line" by a control unit detecting pressure drops in the system. Pressure rises in the system result in pumps being taken off-line by the operation of relief valves. An increase in hydraulic demand in the mid-1970s was brought about by the replacement of all manual 12" presses by the 1400 type press. This was caused by the press which, besides the demand of approximately 1 gallon of hydraulic oil per press cycle, also operated shorter press cycles than its predecessors. Hydraulic saving equipment, which was installed on the majority [and eventually all] of these presses, overcame the space availability problems that the provision of additional pumps would have created.

With the soluble oil hydraulic system at the Blyth Road site, system and equipment leaks quickly evaporated away leaving just a lightly oil-stained area. The non-soluble oil system at the new site offered considerable benefits in terms of reduced wear on plant and equipment, but system leakages were a major source of concern. Frequent checks of all trench pipe-work were implemented, and drip trays were fitted to all presses to contain the problem.

Considerable quantities of compressed air at a nominal pressure of 100 lbs per square inch are consumed by the pressroom, and many other areas across the site. To meet these demands, three Weir non-lubricated air compressors, each of 1000 cubic feet per minute capacity, were provided. From the mid-1980s, these operated together with refrigerant air dryers, bringing further operating economies, including a reduction in the maintenance costs of parts and labour incurred by thrice yearly filter element changes.

To equip the central vacuum system six Lacy Hulbert vacuum pumps were installed, and were capable of meeting all production demands. They proved trouble free during their twenty-plus years of use, and maintained the central vacuum system at the required 25 ins. Hg.

The power house is also equipped with a demineralisation plant that has had only one change of ion exchange resin in 20 years! Up to two thousand gallons of pure water were supplied to the matrix department each day during periods of peak demand. This plant was used to meet the particularly high quality water requirements of the matrix process.

Two stand-by generators were [eventually!] installed at Hayes. The first, in 1972, was a "Waukesha V12" with an output of 1000KVa. Operating on natural gas, it was later joined by a smaller unit from the same manufacturer with an output of 250KVa. Both proved themselves during the miners strike in the early 1970s, which resulted in the three-day working week. During this period the site was allocated a maximum load of 1000KVa for essential services by the electricity board. The use of these generators allowed production to continue, albeit on a much reduced scale. Electrical services to the matrix department, toolroom and distribution building were maintained by three portable generators, loaned to EMI by jacket manufacturers Garrod & Lofthouse for use on the days when Garrod and Lofthouses' electrical supplies were normal. These generators covered many miles during the strike period, travelling back and forth between Hayes and Crawley! The arrangement worked well due to our 3 days' supply not coinciding with theirs! A rota of key personnel, working in pairs, was established for the duration of these restrictive workings, and their task was either to start or to off-load the generators and other main switch-gear during the night. The distribution building was later equipped with its own 500KVa Dorman diesel-oil powered generator.

Two Carrier/Carlyle 16JB Lithium Bromine machines were originally installed to provide air conditioning chillers, and were capable of reducing 1100 gals/min. of water from 80 to 60 degrees Fahrenheit. Utilising 10 lbs/sq.in. steam, these units progressively became less efficient. In 1986 they were replaced by compression refrigeration equipment, in the form of two McQuay ESHT280 units fitted with gas condenser towers. These proved to be far less labour intensive and cheaper to operate, and remain in use today.

Also under the control of the power house is the town-water storage tank. This large, cylindrical tank is sited in the south easterly corner of the manufacturing site. It is 20 feet in diameter and over 55 feet tall, having a nominal capacity of 100,000 gallons of town water, and is partitioned such that the upper quarter supplies the needs of the matrix department. Under high output conditions, up to 3,000 gallons per hour were required by the power house, and the tank provided many hours of reserve on the few occasions when the areas town-water supply was interrupted.

CHAPTER THREE

The Materials

INTRODUCTION

The technology of the materials and processes of record manufacture required constant improvement in order to match the changes in equipment and the demands for higher quality and better efficiency. Over the years the works laboratories of Blyth Road were responsible for both material and process control development.

The laboratories were eventually amalgamated under a single responsibility, and based on the fifth floor of Schoenberg House in Trevor Road, although they retained small operations in

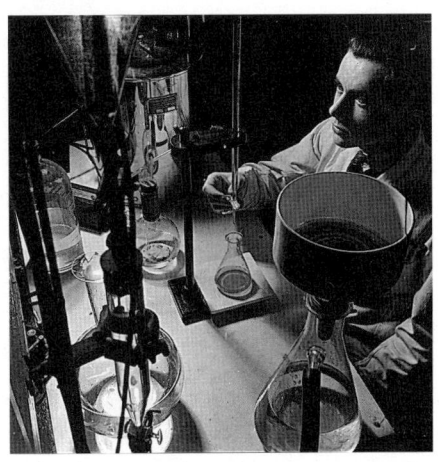

The Works Laboratory.

the matrix, weighing and grinding and pressroom areas. These outposts were primarily used as pilot plants for evaluating new materials and processes.

The laboratory, offically named the record production and development department, was subdivided into five sections;

> Vinyl development
> Paper laboratory
> Microscopy
> Audio
> Chemistry

When the production facility moved to the Uxbridge Road site, a new modern and fully equipped laboratory was built. It was led by Bill Soby, with Jim Hughes as supervisor, and Robin Allen, Paddy Marchant, Ray Saunders, Chris Lawes and John Garodi as section leaders. The laboratory continued to refine and develop all the new raw materials involved in disc manufacture, documenting the materials and the processes into the Manufacturing Standards Manual. This manual was sent to all overseas factories, and was regarded as a technological lifeline for them to Hayes, the mother plant.

In the halcyon days of vinyl, 1972-1978, the laboratory employed 25 staff. In 1980, however, it was recognised that vinyl materials and processes had reached a plateau, and the laboratory was disbanded. Key technologists were assigned to the appropriate production departments.

RECORD COMPOUNDS

In 1908, the records were made from a material known as shellac, which is a natural thermoplastic resin derived from the secretions of the Lac beetle. The female of the species secretes the resin so as to bond herself to the bark of the tree from which she derives her sustenance. The Lac beetle is a native of the Indian sub-continent, but is also found in Thailand and Brazil.

The shellac compound used for 78 rpm records was a mixture of 65% slate powder, up to 27% shellac resin, and small amounts of manilla copal, wood resin, carnauba wax and carbon black.

The marked 7" vinyl biscuit which will be broken into manageable slabs after being air cooled.

The slate powder was delivered to the Hayes site by Great Western Railway wagons, straight into the factory sidings, where it was then handled by the factory's own engine.

The powder mixture was processed through a two-roll dispersion mill, and a sheeter mill, to produce the "biscuits" of material ready for use. This equipment was later used in the early processing of vinyl material.

Supplies of this material were generally not difficult to obtain, except during the Second World War. Records were still being produced using shellac as the base constituent, which had to be imported from, among other countries, India. Initially it was felt that records would play a very small part in the war effort, but it soon became obvious that sales were increasing and the stock piles of materials were rapidly being depleted.

The Hayes factory had to tackle the problem of material shortages, if there was to be a continual supply of music. First experiments were conducted which incorporated a filler to replace some of the shellac. Virgin material was used for the playing surfaces, while the interior of the record was literally filled with any available scrap material, even sweepings off the floor. This proved relatively satisfactory, but as time progressed the scrap content tended to migrate from the inside of the disc to the playing surfaces, causing considerable surface noise.

Wear testing 78 rpm Shellac Records.

To stop this happening extremely thin metal plates were added between the filler material in the middle, and the virgin material on the playing surfaces. Technically this was now a sandwich and the question was asked, could anything other than scrap be used to fill it?

The answer was yes, and before long someone suggested that maps on silk material, or money, would do, and if the metal was magnetised a simple compass could be constructed. Whatever for, one may ask?

Records were being sent in Red Cross parcels to prisoner of war camps, and it was felt that such contents could be very useful to potential escapees. This may sound far fetched, but next time you are in a book shop, look out for the original edition of the Colditz Story, in which you will see photographs of records broken open to show their contents!

Another scheme to help the war effort was the recovery of unwanted 78s, which were mixed with a small quantity of virgin material to extend supplies. Dealers throughout the country were encouraged, by The Gramophone Company's magazine, "The Voice", to run advertising schemes asking the public to return unwanted records. The results were most gratifying, with a dealer in Plymouth reporting that 30,000 discs had been returned to him alone. This scheme undoubtedly helped to ensure that records were made available to customers throughout the war.

The plastic used today for the manufacture of LPs and singles is affectionately known as "vinyl". Vinyl, however, is a shortened version of polyvinyl chloride, which itself is still not an accurate description of the material. The plastic we know as vinyl is actually monomers of vinyl chloride and vinyl acetate, polymerised together to form a copolymer of polyvinyl chloride polyvinyl acetate.

The word polymer comes from the Greek "poly", meaning many, and "mer", meaning the same; that is, many of the same molecules joined together. For the sake of simplicity we have used the abbreviation P.V.C., when describing material elsewhere in this book.

For the last 25 years the supply of copolymer has been relatively stable and uninterrupted. The principal suppliers being The European Vinyls Corporation, a subsidiary of ICI Chemicals, and Ato Chem of France. Other suppliers have included Wacker Chemie of Germany and BP Chemicals.

Physically the copolymer is a white free flowing powder, and from the mid 1960s until June 1992 was supplied by road tanker and stored in silos holding up to 100 tonnes. At its peak the factory was using over 5,000 tonnes per year.

The copolymer on its own, of course, is not suitable for record production. It also needs a thermal stabiliser, a lubricant and a colouring pigment. Copolymer, if processed without pigment, would produce clear records, not black ones. The pigment, carbon black, was used only to colour the vinyl; it had no other effect, and it is significant that a mere 0.3% of carbon black is needed to give the pressing its jet black appearance!

The thermal stabiliser is needed to prevent the formation of harmful decomposition by products, while the P.V.C. is heated. Early stabilisers were based on lead compounds and, as lead was recognised as environmentally unacceptable, it was replaced by barium and magnesium compounds. The lubricant, a wax-based material, was added in small amounts to give the record an extended playing life.

In 1992 the major reduction in the demand for vinyl discs required a rationalisation of the material and its associated mixing costs. The vinyl today is supplied by Ato Chem of France, as a compound ready to be fed directly to the presses.

The development of the materials used in the manufacture of the modern records can be traced back to the early 1950s. In preparation for the production of the new long-playing records, trials were conducted in the latter part of 1951 with "Geon", a vinylidene chloride/acetate co-polymer supplied by British Polymer and used, then, by Decca Records.

This did not prove satisfactory, however, and therefore initial production at Hayes utilised the RCA Victor formulation. This was shipped from the United States, and comprised:

VYHH [Vinyl chloride/acetate co-polymer]	97.0%
Carbon black	1.5%
Di-basic lead stearate	1.5%

This remained the reference material, and eventually two British polymer manufacturers became involved as suppliers to EMI. One was Bakelite of Aycliffe, County Durham, makers of "Vinylite" trade mark polymers in the United Kingdom, who developed an alternative material designated VYSL, which was subsequently used in quantity. The other was ICI Ltd. of Hillhouse in Lancashire. ICI were not able to offer a suspension polymer with VYHH characteristics, but went on to develop an emulsion polymer, known as Q43/60, which was used in place of VYHH in the material formulation for many years.

Although initial supplies were imported from American sources, ICI became the main supplier of P.V.C. to EMI. The association of EMI and ICI pioneered the early development of vinyl, and ICI continued as the sole supplier to EMI, Hayes, until 1978, when market forces dictated that other suppliers be used. These other suppliers included BP [U.K.] in 1979, Ato Chem [France] from 1980, and, occasionally, Wacker Chemie [West Germany]. Ato Chem also supplied all of EMI's European plants except EMI Electrola in Germany, who were supplied by Wacker Chemie.

In the early 1960s, attempts were made to introduce surfactants to the formulation to make the P.V.C. antistatic, a very desirable property. The problem, as always, was the incompatible nature of the antistatic agent with quality pressings, causing undesirable aural effects on the disc, particularly at the end of a side, known as "end-line build-up". The idea was eventually abandoned in the U.K. in the mid 1970s, but continued to be investigated by Toshiba-EMI until the demise of vinyl in Japan in the late 1980s.

Until the early 1960s the formulation had remained relatively simple, copolymer, di-basic lead stearate, wax and carbon black pigment, and the material could cope with press cycle times that were more than 50 seconds per 12" disc. Raw materials prices in Japan in 1960s were very high, and in an attempt to reduce the cost, a filler polymer, which was much cheaper, was introduced into the formulation.

At Hayes this idea was quickly adopted and important beneficial side effects were immediately recognised. The filler polymer increased the apparent melt viscosity, enabling the press cycle times to be reduced by as much as 20%, the only detrimental effect being a marginal increase in surface noise. The filler polymer was a newly introduced ICI homopolymer of P.V.C. having a high molecular weight. At the processing temperature used, the filler polymer does not flux or melt, but remains as discrete particles.

As recordings and replay technology advanced, the surface noise generated by the homopolymer became unacceptable, and it was replaced in 1972 by a medium acetate low molecular weight copolymer. This behaved in a similar way to the homopolymer, but was softer, and did not affect the surface noise.

In 1974 the "oil crisis" was upon us, and P.V.C. prices soared. The main task for the factory was to maintain scarce supplies and to contain spiralling material costs. Many experiments were tried with a variety of materials, including extender resins, to overcome the shortages.

The most successful method of securing P.V.C. supplies, however, was to extend the addition of homopolymer from 10% of the formulation to 50%. Homopolymer was more readily available than copolymer and was less expensive. This type of formulation was widely used in the industry but in quality terms was generally unsuccessful at Hayes.

The feed aperture of the Banbury Mixer at the mezzanine floor level.

Up to the mid 1960s, the processing equipment used to convert the copolymer and additives to a plastic melt was a conventional Banbury intermix with roller mills.

The equipment was large: the Banbury stood 20 feet high, and was fed manually with the required proportions of copolymer and additives. The ingredients were roughly mixed in this equipment, and gelled into a crude plastic melt. The roller mills were approximately 4 feet in length, and 2 feet in diameter, their purpose was to mix intimately, and disperse, the crude melt from the Banbury equipment. These mills produced a flat uniform continuous sheet of hot plastic, which was then cooled and cut into the required biscuit size. As the production process evolved, the sheets of plastic were later fed to a "chopping" machine, that cut the sheet into small cubes, which were then "bagged" and stored for future use.

The whole process was housed in a department known as "the black room", a name that was inherited from the shellac processing plant. The name survived through the years, even applying, although inappropriately, to the modern automatic mixing plant, used until 1992.

The operator, Len Taylor, wears ear defenders as protection against the extremely noisy dicing machine used to produce cube-cut granules.

Automation brought many changes in both the physical nature and the handling methods utilised for the P.V.C.

The first major development in the evolution of the use of P.V.C. began in the 1960s, when a new highly advanced compounding extruder was designed and supplied by Werner and Pfleiderer of West Germany. The novel feature of this machine was that each press could have its own extruder. This could be fed with powder material or, as it is better known, dry blend.

The design of this extruder, affectionately known henceforth as a W & P, was so advanced that the quality of the extruded plastic was comparable to the milled material. The obvious economic advantages were that the expensive black room equipment was obsolete, and could now be replaced with an efficient automatic weighing and high speed mixing system. This system fed mixed material to 60 tonne silos, which in turn fed into an automatic air-veying system connected to each extruder. The old mill and sheeting equipment were scrapped.

With the high speed automatic 12" presses being developed, the output of the W & P extruder was not going to be able to keep up with them. The copolymer used at that time was a porous particle polymer. The limiting factor of the extruder output was the density of the copolymer. The answer was provided by ICI, who, over a short time, produced an identical polymer, but with a spherical shaped particle. The density of the powder increased from 600 to 800 grammes per litre, allowing the W & P to produce ample output for the presses. This polymer has remained unchanged to the present day, and was subsequently copied throughout the industry.

The additives used in the manufacture of the compound also remained largely unchanged, except one. The lead based stabilisers used in early formulations were replaced by barium/cadmium soaps, when lead was recognised to be toxic. The heavy metal cadmium was, in turn, listed as unacceptably toxic, and was finally replaced by a magnesium/zinc based stabiliser.

It is interesting to note, however, that the desire to have black records has ensured that carbon black, which gives records their colour, was used in both the first shellac records made at Hayes, and continues in use today.

RECORD LABELS

Paper as a major raw material in disc record production comes second only to vinyl, in terms of expenditure.

There are two main areas wholly dependent upon paper that have traditionally formed an integral part of the gramophone record; these are the record label, and the packaging, which is described in the next section.

The most obvious manifestation of information about any record is the label at its centre, identifying programme and artiste as well as the record company, copyright data and a catalogue number.

Labels on the earliest 7" records were hand-written affairs etched into the metalwork from which the record was moulded. Paper labels came in around the turn of the century, and fairly soon became the standard for shellac and, later, vinyl records. The suitability of paper labels becomes apparent when it is seen that the first shellac records made at The Gramophone Company Ltd.'s factory at Hayes in 1908, and the 12" microgroove vinyl records from automated presses at the Uxbridge Road plant 84 years later, all used them.

Little information on the early use of paper and printing inks is available today. It is most likely that all that was required was for labels to be of sufficient strength to survive the moulding process, and for them to be good enough in appearance to be acceptable commercially.

By about 1950, after the shortages of wartime, the situation had stabilised, with labels being supplied to EMI at Hayes by Messrs. Harrisons, from their plant situated conveniently close to EMI's factory.

In-house label printing at Uxbridge Road with Dick Hares and Roger Aldworth apparently oblivious to the camera!

This supply was supplemented by "in house" printing and manufacture, which provided a source of fully finished labels under our direct control. This facilitated rapid turn-round of orders, particularly during periods of high or fluctuating production.

Printing was still almost exclusively by letterpress, with font set by hand, as was customary then. The labels were printed on a reel, then punched into shape, and lastly had the centre hole drilled, all as separate operations. Actual designs tended to be standardised, and apart from the prestigious HMV label, were generally kept simple, often being made from different coloured friction glazed papers titled in 'gold'. This gold effect was achieved by overprinting the label information in a clear medium, and then dusting it with bronze powder while the medium was still "tacky".

To make the records, labels were fed into moulds by hand, together with heat softened shellac stock, and pressed between chromium faced stampers. When problems with labels did occur they were mostly dealt with by on-site production personnel being able to recognise the nature of the problem, and draw upon their previous experience to effect a solution.

This May 1952 photograph shows the 12" LP pilot pressroom equipped with twelve record presses.

The first major change in label technology was bought about to meet the release of EMI long play records in October 1952.

Manufacturing standards for long play records came from the U.S.A., where vinyl microgroove records had been launched on to the market four years previously. At Hayes, twelve modified, manually operated, semi-automatic presses met the initial demands. As with shellac moulding, labels still had to be dried before use, and problems were likely to occur from the drying conditions, and also the variables encountered in subsequent processing. This was especially true as they were using non-automatic production machinery. A lack of proper understanding of drying principles, with such ostensibly minor differences in press cycle time, rate of ram rise, temperature and pressure could all affect label performance.

The main change that the American standards called for in microgroove record labels was the use of higher strength body paper to meet increased shear forces encountered in moulding vinyl, compared with the softer shellac compounds.

For the 7" 45 rpm record, originally designed to replace the 78 rpm disc, labels were required to have a high friction "nonslip" surface. This was intended to reduce rotational slippage when playing these lightweight and large centre hole records on rapid auto-change machines [juke boxes]. The high frictional coefficient was achieved by coating the surface of the paper with a slurry, heavily loaded with finely powdered barytes.

Such changes in the paper and process enabled a start to be made, but with the projected increase in vinyl record production it became essential to gain a deeper understanding of label technology and the causes of failure.

From these beginnings a long-term programme was planned to improve and rationalise performance. An important part of this programme was to establish optimum drying techniques, to minimise problems from ink sticking to the stamper in the press, and then later to develop procedures for handling labels on autopresses.

After determining these objectives, a systematic investigation of the causes and effects of label failure was undertaken, to meet current production needs, as well as becoming part of the autopress engineering development programme. Simply solving day-to-day problems was not enough, as rapidly evolving automation called for a more fundamental approach. In consequence of this, a working group was formed comprising technical staff from the engineering, laboratory and production departments.

A study of the conditions that affect label performance was made, and as understanding grew so technical expertise widened to include paper makers and coaters, specialist ink manufacturers and printers.

As a result label development kept pace with engineering advances and this eventually led to record labels essentially different from those used in the rest of the industry. This, together with a completely changed method of drying, made them well suited to moulding on automatic presses and to meet the predicted increases in market requirement for vinyl discs.

At first these new labels were produced specially for EMI's own use but, as the advantages over existing labels became obvious, the body paper on which they were based gained wider acceptance. Eventually it became established as the normal record label paper made by major paper coaters supplying the industry. This helped to uphold quality standards as its use widened to nearly all record manufacturers.

The most important characteristics built into the new label were as follows:

A long fibre, high burst body paper that still retained good strength after drying.

Paper with a balanced coating of casein bound china clay on both sides to prevent curl during drying, so as to present a flat label to the press.

The coating on the reverse side of the label also sealed the paper to stop "vinyl bleed" through to the front face. This is a phenomena where the label colour becomes patchy because the vinyl enters the label quite deeply in places, and can be seen through it.

Good coating opacity, from dense pigments, reduced discoloration after moulding.

The incorporation of a thermoplastic polymer into the slurry provided flexibility, and hence avoided the coating cracking when the label was punched, and crazing when it was dried.

Calcium stearate in the coating provided lubrication to assist in cutting the paper in bulk, and also when punching and drilling the centre hole.

Low air permeability, needed for reliable suction pick up of the labels from the press magazine, was also achieved by the above coating system.

An added advantage was that coloured pigments could be incorporated into the basic white coating of the paper, giving a label with a ready-made coloured background, thus avoiding a further printing operation. Labels of this type pressed well as there was very little ink on them, and a whole range of different coloured backgrounds were produced. These were ideal for EMI to do just the overprinting of title information on site. This was actually done for several years, within the matrix and label store department.

As for the actual inks used in printing the labels, changes were made initially after development work by one specialist ink manufacturer produced a successful ink, but the colour range was restricted and supplies were limited.

A "successful" ink was one that printed well, had good colour heat stability and did not stick to the stamper during moulding. Mostly this was arrived at by trial and error testing, but in time a working list of approved inks suitable for autopressing was compiled to be used by label printers.

Printing by the early 1960s had changed to offset lithography, often by the four colour process suited to the greater range of dedicated designs that came into fashion in the 1960s. This worked to EMI's advantage as the ink films on the label were much thinner than those used by the older letterpress method, and further reduced the risk of sticking problems.

The method of drying labels was changed drastically from the established practice of drying them "en bloc" in batch ovens, until the

Labels emerge from the drying oven after approximately an hour and Roy Kerr prepares to remove them.

inks pressed satisfactorily. This normally took several hours, but sometimes as long as two or three days were spent with "difficult" labels. After looking at infra red, microwave and other newly available techniques, the method finally adopted was developed from the study made to understand the principles of drying and establish optimum drying techniques. It consisted of a circulatory air oven, where labels, supported on rods through their centre holes, were dried by conveying them though the oven chamber. This gave unrestricted hot air access to the entire surface of each label, reducing the drying time to about an hour. After this they were sealed into polythene bags to keep dry.

As thermal energy is responsible for both the physical and chemical changes of the label, much thought went into choosing the temperature to be used. The objective was to obtain maximum effective drying in the minimum time, taking into account the interrelated conditions which react on each other, and their effect on deterioration in the papers and inks comprising the label.

Moisture in the label was a major cause of pressing problems, and a temperature of 165°C for 1 hour was selected as optimum to remove both the absorbed, and some of the molecular, water from the cellulose fibres comprising the body paper. This had the effect of reducing the rate of subsequent moisture pick up when dried labels were again exposed to the pressroom atmosphere. With good hot air flow over the surface of each label the printed ink films were also hardened through oxidation, providing additional security in moulding.

By 1972, when the automated vinyl plant came into operation, labels that had been developed for autopressing were already well established and became part of the new production system with two conveyor ovens for drying them.

Although record rejection rate attributable to 'label faults' had fallen following more extensive use of the new labels, problems still happened from time to time. They were usually of a localised nature, and could always be attributed to some malfunction either of the paper, the ink manufacturers or the printers, or because of incorrect plant operation or simple non-adherence to procedures. Confidence in the principles established earlier remained, as it was always possible to resolve such problems by returning to the original material specifications or operational procedures.

There was still, however, a need for continuing development, refining different parts of the basic concept in the light of practical experience. This led to further changes, which included:

> The introduction of high density polythene film bags, replacing low density ones, for packing labels after drying, enabled them to be packed straight from the oven without waiting to cool. When

Bags of dried labels being vacuum sealed.

followed by vacuum sealing it produced compact units suitable for storage, until required for use, as ready dried labels.

Lathe turning of labels in bulk gave a cut edge of more consistent quality freer from loose debris than the conventional punching technique. This method, adopted by EMI's major U.K. label supplier, was based on the technique for producing labels used at EMI Electrola, Cologne.

A surfactant added to the coating of labels for 7" records gave a higher electro-conductive paper reducing static charges, which lowered the risk of multiple label pick up in the press.

A tool was designed to "stipple" the label area of positives to produce stampers covered with a mass of small tetrahedrons about 0.002" high at a density of 100 per sq. cm. The stipple replaced the less controlled practice of scuffing stamper centres. By indenting the labels as the press closed, the stipples held them firmly in place, while the plastic flowed outwards to fill the mould cavity, thus preventing radial burst.

Paper labels had survived during most of the lifespan of the record, whether it be shellac or vinyl, at least as far as mainstream production was concerned.

These, in outline, were the major changes in labels and the way they were processed that enabled them to be used in their millions at

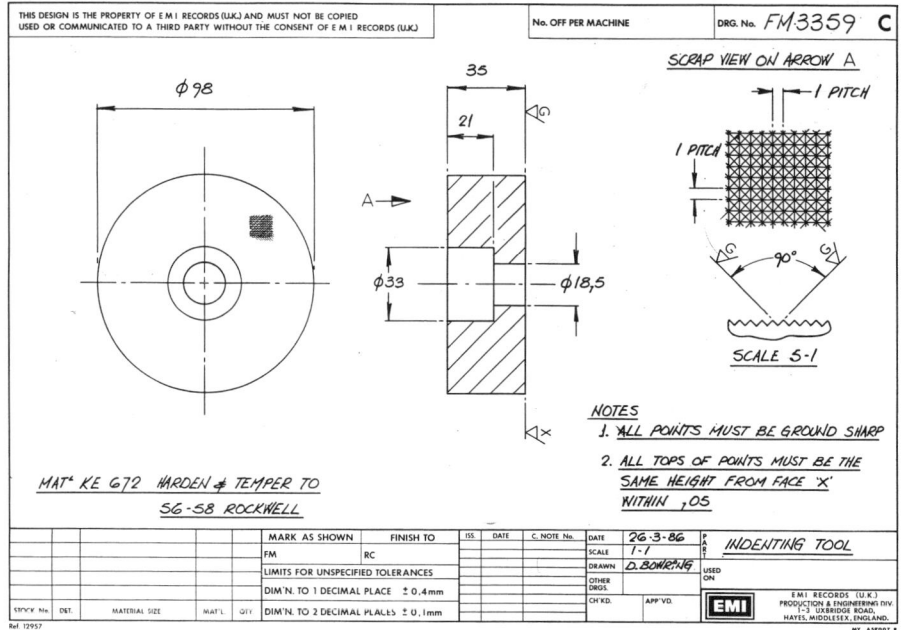

The indenting tool as used in a hydraulic press.

Uxbridge Road, at one time the largest facility for manufacturing vinyl discs in Europe.

Only in the last decade or so has the supremacy of paper labels been challenged. This was when the alternative of paperless labels, inked at the press, became the mainstay of 7" record production. EMI introduced these following their development on the continent, at Pathe, and EMI's own initial trials in 1964. Thus we come back to where we started before paper labels came into being. As with the earliest labels in the 1890s, these, too, are etched directly into the metalwork, and even the size of the record is the same!

RECORD PACKAGING

Almost from the beginning of the life of records some form of packaging had been part of the commercial presentation of the product. In the early days, however, the novelty was in having sound actually coming from the disc. Scant attention was paid to how it was packaged, so a simple paper bag sufficed, and eventually became accepted as normal. Later, some bags were printed with a Company Trade Mark, others with pictures of artistes and lists of their records, and still others with pictures of the types of gramophone then available. The information printed on sleeves today, and taken so much for granted, was in those days reserved for bulky sets, which were graced with analytical notes printed on a loose insert or booklet.

Perhaps it was the inadequacy of the paper used to protect the brittle discs from breaking that produced a technical awareness, which in turn led to the introduction of bags made from heavy duty cardboard. These became very fashionable during the 1920s and 1930s, and were supplied by dealers who had their names printed on the face. They formed a permanent advertisement of who had supplied the record, and could supply others!

Such durable packs ended with the war, when bags reverted to paper, although this, too, was a restricted commodity, and with a variable and generally lower quality than in the pre-war days. This situation continued for about a decade with some of the worst paper getting into the system as late as 1950, causing extensive damage to records. This came from chalky inclusions in the paper stock from which the bags were made.

The identification of this problem caused a new consciousness of the relationship between records and their packaging and the effect of one on the other. With the return of a quality choice of paper the previous "Standard Record Bag Paper" was upgraded to a superior and more expensive brown Kraft paper. This was a distinct improvement and resolved the damage problem, but the record pack remained what it always had been, a brown paper bag with holes in it to read the labels of the record inside. It was not until the approach of vinyl that things really changed.

Whereas the paper labels for vinyl discs evolved from those used for shellac, packaging for the microgroove record was a completely new feature, and not simply a case of refining existing bags in some way or other. For one thing the marketing standards designed for long play in the U.S.A. were different, and more akin to those of album sets of 78 rpm records. They had programme notes for classical issues and "track listings" for others. It was also necessary to take additional technical factors into account; as compared with shellac, vinyl was 'soft' and its close-cut microgroove was more vulnerable to damage, so record protection assumed new importance.

American packaging manufacturing standards, which formed part of the total vinyl product, called for the bagged disc to be contained within an outer jacket, printed with artwork and notes related to the recorded work.

The change was a complete departure from all that had gone before. In The Gramophone Company factory at Hayes the 78 rpm records in their brown paper bags contrasted drably to the new vinyl disc, packed in its polythene inner bag, and enclosed within a multi-coloured outer! There was simply nothing to compare with the particularly resplendent classical HMV jacket sporting a handsome full colour reproduction of Barraud's 'His Masters Voice' painting, framed with gold scrollwork and emblazoned with Coats of Arms by Royal Appointment. Complete with polythene inner this empty package cost 'half a crown' [12.5p], in 1952 a considerable sum.

As at first vinyl records were both made and packed by hand, the initial requirement was simply to obtain packaging that met the American Standards. The protective polythene inner did not present a problem. The American jackets, however, were made from a core of thick ticket board, with a pictorial label stuck over the whole of the front face, and a liner with printed text stuck over the back. No British manufacturer was set up to make this type of pack, and, anyhow, the 'jumbo' rolls of ticket board used were not available in this country.

Other than importing fully finished jackets, it therefore became necessary to follow established principles rather than the American practice. By 1952 the microgroove record had already been made in the U.K. for two years, by EMI's competitors, and there was already a scattering of small printers set up in business to make record jackets. Most of their output was taken up elsewhere. So, to meet EMI's LP launch date, and subsequent releases, initial supplies of jackets, apart from the deluxe HMV classical, were made by Messrs. Harrisons Ltd., our main source of 78 rpm record bags. The early jackets were made from 0.015" board, edge bound with cambric tape, and printed in a common design for each issuing company within the group. They were then overprinted with title information and appropriate notes on the reverse.

With the expansion of the LP catalogue, however, a certain monotony in appearance of the EMI jackets was becoming obvious, and a move was made towards individual artwork for each catalogue item. Fortunately, by this time other print houses were entering the field to meet increasing demands for the new style of jacket. In this developing situation it was necessary to acquire increased technical understanding of the bag and the jacket. As with record labels, a study was undertaken, with the object of arriving at optimum technical design for record jackets and inner bags. Expertise on jacket boards and their fabrication came from Messrs. Garrod & Lofthouse Ltd., and on inner bags from Harrisons Ltd., while EMI contributed its own specialised knowledge on the product to be packaged.

Although primarily concentrating on the record jacket it soon became necessary to widen the scope of investigation, linking together the record, the inner bag, jacket and transit box as an integrated whole. Some of the main packaging features arising from this investigation included:

The finished jacket was designed to have good balanced structural strength, with in-built dimensional stability, so that it did not itself deform, and thus helped to keep the enclosed record flat.

For the inner bag, low density polythene film was used, contoured to the shape of the record, and later as a loose lining in paper bags held by spot welding or a tacky adhesive. Unlined paper bags were then used only for some lower priced records.

For the packaging box to transport and store jacketed records in bulk, a change was made from stapling and assembling boxes and lids on site, which had begun during the shellac period. The new box was based on a Capitol U.S.A. design, consisting of a die cut and creased corrugated board, which could be rapidly folded into an interlocking one piece box and lid unit, without resorting to stapling.

The study also embraced techniques of disc handling from the viewpoint of both the manufacturer and the consumer, and the effects of environmental factors on the record and its packaging; in the warehouse, during transportation, and longer-term while in the possession of the customer. The recommended practice of vertical storage of jacketed records to minimise warpage was established during this period, a preference that remains to this day.

Results from these investigations were to prove invaluable when the change was made to autopress moulding, with its associated packing techniques. Meanwhile they helped to resolve packaging problems as they occurred. These included record warpage and jacket edge rupture. Other issues included the problem of the substances used in adhesives and the polythene film attacking the record during undisturbed storage, and even the growth of fungi under certain conditions!

By 1969 the evolution from hand to semi-automatic packing had occurred. Records still came from the press un-bagged, stacked horizontally in transit boxes with rigid separator discs after every fifth record, to help keep them flat until they had hardened sufficiently for inspection and packing. After audio approval, on a batch sampling basis, every record was visually inspected and those with moulding faults, or that had become damaged, were discarded.

Acceptable records were hand fed into packing machines developed at Hayes, which both bagged and inserted them into outer jackets. Variables accommodated within a hand packing process could now no longer be tolerated. This led to more stringent dimensional limits for packaging, as well as the need to specify requirements for the air permeability and smoothness of inner bags that were now picked up by suction.

The ease of inserting the record into its bag, and the bagged record into the jacket, had to be improved and made more consistent. This was achieved by changing to high density polythene film for the inner bag lining. The introduction of this film also greatly reduced the risk of additives from the film migrating on to the record surface. Other changes included modifying the jacket structure, involving the incorporation of box spines and by increasing the overall length and width.

With a bright future looming for the microgroove record, a separate ambitious development was being undertaken for Garrod & Lofthouse, who had become EMI's major supplier of jackets. This was a machine that inspected the record before bagging, and constructed a box-like jacket around it. The open structure of the pack, however, did not allow sufficient support for the freshly moulded record, which tended to warp during storage inside the pack, and the development was not progressed further.

It was not until the mid 1970s, after the vinyl plant had become established at Uxbridge Road, that the next packaging change was made. EMI long play records had been subject to an increasing number of unfavourable comments, from the trade and in consumer journals, for crackle and surface noise. Record returns analyses from major dealers suggested the main cause of complaint to be damage, in various forms, to the surface of the disc. Extensive on-site surveys showed that records came off presses in good condition, but were damaged in the process of conveying boxes of un-bagged records and separators to the examination department. Here also additional surface scuffing could occur as records were fed into the mouth of the packing machines.

An immediate modification to the profile of the separators inserted every five records [changing from round ones to square ones with chamfered corners] overcame the conveying damage problem. The real solution, however, was to capitalize on the reliability of the 1400 press and bag the record immediately it left the mould, for which purpose bagging units were fitted to the presses. These did away with the handling of un-bagged records in bulk, and visual quality could be controlled by systematic spot checks at the point of manufacture in the pressroom, and in the examination department.

Conditions were now set for the next move to increase throughput and reduce costs by installing two Winkler & Dunnebeir high speed jacketing machines. These were designed to handle the long production runs met with during the years when LP production was at its peak. On these machines, magazines were loaded with pre-bagged records, which each machine could jacket at rates of up to 3,000 per hour, in either standard jackets or gatefolds. Such speeds meant a further upgrading of quality specifications for jackets. To find room for the quantities of sleeves now involved, a temperature and humidity controlled high bay storage unit was built at Uxbridge Road with an initial capacity of 14 million empty packaging items. This was

The Winkler & Dunnebeir high speed jacketing machine seen packing the 10th Anniversary Celebration Record in 1982.

constructed to meet future production demands, and also helped to reduce the occurrence of jacket problems. Jackets themselves were delivered from manufacturers in predetermined quantities, packed in boxes already stencilled with contents information, which were subsequently used for the despatch of finished product to the distribution centre and the shops.

Short manufacturing runs continued to be jacketed by hand, but by the late 1980s LP record production had fallen to a new low level. Although the effect of this was offset somewhat by the growth of the 12" 45 rpm record, eventually long runs became less frequent. With a multiplicity of smaller orders the W & D machines were replaced by ergonomically designed hand packing stations, and a return to the methods of earlier times, albeit with much improved efficiency.

1952 saw the launch of EMI's 7" 45 rpm disc, which, although presented as an alternative to the shellac record, has survived through four decades, doubling its' playing time for a while as an "Extended Play" version, which existed until about 1970.

Packaging for 7" product was modelled on that for the shellac record, and consisted of a simple paper bag with two centre holes. Technically it offered few problems, as the damage risk was small because the record profile itself gave some protection, and high modulation tended to mask the aural effects of surface scuffing. Discs were packed by hand in white Kraft paper bags, flexo printed in the reel, with selected logos as required.

Bags of this kind have remained in use to the present day, with only few modifications. The most important change was the dimensional revision in the 1970s, to meet the requirements of the auto-bagging units fitted to the Windsor SP130 injection moulding machines. A further change which has persisted is the fashion for black bags, instead of white. This was at first achieved by pigmenting paper with carbon black, and later by printing with black ink on white paper, which handled more satisfactorily on the press because of its lower and more consistent air permeability.

The dedicated 7" bag was originally used for Extended Play records, and during their decline in popularity was revived on a more enterprising scale for 7" single play records. It was subject to constructional constraints for use on bagging units and in appearance looked like a miniature LP jacket.

So did the bags for 12" 45 rpm records, introduced in 1979 as a big brother to the 7" record. These discs started life as limited edition collector's items. With the aural impact of their even higher recording levels, their ability to make available extended versions of the recorded programme, and their being packed in bags featuring individual designs, they rapidly became popular and formed a significant part of pressroom output during the 1980s. Made to a similar specification as existing special inner bags for LP records, and with a low friction varnish coating they bagged well at the press. Their fabrication, without box spines, helped to keep packed discs flat immediately after moulding and eventually this enabled these records to be bagged, boxed and shipped direct from press to warehouse.

Changes throughout a long time span are inevitable, caused either by an established raw material becoming unavailable, or, more positively, by the results of development. Record packaging was no exception. In particular it was affected by developments initiated by EMI, and in addition to those changes already referred to, there are four others of note:

> The theoretical preference for inner bag paper is to have a smooth surface in contact with the disc. In practice this often gave problems inserting the records into the bags, both by hand and on press, due to electrostatic charges and particularly under low humidity conditions. For many years paper with a soft, rough inner surface was used satisfactorily to ease the situation. On the introduction of Direct Metal Mastering in 1982 with its close shallow groove, superficial damage to land bordering the groove

caused aural crackle, so a new smoother paper was sought. Auto-bagging reliability with the new paper remained a problem that was finally resolved by using polythene lined inner bags.

The importance of the paper grain direction was such that direction control in bag make-up was necessary to ensure correct opening on press bagging machines. This was specified as the grain running from the mouth to the base of the bag.

The problem of aggressive substances migrating from packaging adhesives and attacking the record surface was eliminated by using hot melt adhesives for jackets, water soluble dextrin for inner bags, and certified safe adhesives for jacket lamination and polythene inners.

Following the period of cellulose acetate film lamination for jackets and brief trials with polypropylene film, the technology of later years saw cost effective ultra-violet cured varnishes become the normal commercial finish.

After the unimaginative brown paper bag of the shellac record, vinyl created a whole new era of packaging to match its times and fashions, even developing into an art form of its own. The vinyl discs themselves have a potentially long life when properly handled and stored. It is here, as well as during the interim period of warehousing and shipping, where they may be subjected to unduly severe conditions, that packaging has made its contribution.

CHAPTER FOUR

Working in the Factories

BLYTH ROAD DURING THE WAR

During the First World War, the production of gramophone records continued in the record factory. In the Second World War, however, the factory was more seriously affected by the conflict, and many changes in the life of the factory and its staff occurred.

One of the most obvious effects of the war was the black-out regulations that were brought into force before the outbreak of hostilities. Strict controls were imposed during the hours of darkness to restrict, and if possible eliminate, light being visible outside the buildings.

Air raid wardens were appointed, from within the staff, to carry out this task. They are, perhaps, best remembered for the cry of "put that light out". What is forgotten is that they stayed on duty above ground during all hostilities to make reports on the raids, and gather what help they could to deal with incendiary fires and casualties. Every building in the Blyth Road complex had its own wardens, who patrolled the roofs watching for incidents and "those wretched lights."

Air raid warnings were an every day experience, with the wardens on the roofs to give warning of impending attacks. Records were kept of the time lost to production by personnel having to take cover during air raids. A comprehensive network of shelters was built throughout the site. These ranged from the famous Anderson type, through to custom built ones, capable of withstanding all but a direct hit.

It is reputed that the German Air Force had instructions not to bomb the Blyth Road complex, as its' distinct shape and location helped them locate their position relative to London. V.1 flying bombs were, however, no respecters of any such orders and the worst single incident of the war at Blyth Road was the explosion of such a weapon on the 7th July, 1944.

It exploded at lunch time on top of the canteen shelter where many workers had taken refuge. The resulting devastation left 37 dead, and 11 injured, some very seriously. The canteen served the whole site and was well used by disc factory staff, this day was no exception, with a number of the dead and wounded coming from the record factory.

As the war progressed so did the number of personnel called to serve in the Forces. This led, as in other industries, to jobs that had long been the exclusive preserve of the male populace being carried out by women. The Gramophone Company, however, was probably affected less than other manufacturing plants. This was because a large number of women were already employed in the examination, edge grinding and finishing departments [although under the control of a male foreman!]

The Company formed a Home Guard Unit, which became a highly trained and capable group. The Home Guard had its own magazine, which was published in the later years of the war.

Was this Home Guard celebration in the Works canteen for Christmas or Victory? Is that young Fred Tiley in the centre background?

During the Second World War the emphasis of the talks and lectures in the canteen changed. The venue played host to many army specialists. They were there to conduct demonstrations and lectures on the strategy to be adopted by the EMI Home Guard platoon, in various situations. The works laboratory, which was located in The Gramophone Company's premises, also gave lectures and demonstrations of the techniques to be adopted in case of gas attacks.

As well as instructions on what to do if attacked, there were also lectures of more aggressive vein. These included lectures on the use of home-made weapons. They were conducted with practical demonstrations being given at the week-ends, in the wide open spaces of Lake Farm.

Other signs in and around the factory that the country was at war included personnel carrying gas masks, and the increased number of bicycles visible at shift change-overs. There was even the sight of the occasional car with a gas bag attached to the roof, which was filled with producer gas used as a substitute for scarce petrol supplies.

Employees whose positions or journey to work involved them having to travel on the railways would have noted a huge increase in goods traffic. They would also see, both prior to and after their journey, huge posters which would ask "IS YOUR JOURNEY REALLY NECESSARY."

The Great Western Railway, in common with other railway regions, realised that this increased goods traffic would need to be regulated in some way if they were to operate effectively. As part of their control of the flow of trains, the construction of a relief "up" goods line next to the factory was soon under way.

The building of this line was followed with interest by the personnel of the matrix department who had a panoramic view. Soon the line was in use, allowing goods trains to back up one behind the other, awaiting clearance into London. A water tower and column were also erected near to the factory, to allow steam engines to replenish their water supplies if facilities nearer the capital had been put out of action.

Security and clocking in and out procedures were strictly enforced. Clock procedures not only allowed the employee to be correctly reimbursed, but also served as an accurate check as to who was actually on site at the time of any incident. This meant that searches could be conducted very accurately. All employees carried a Company-issued identity pass giving their personal information plus the areas of the site they were allowed to enter, and these were strictly enforced. The national identity cards, which everyone had to carry, supplemented them.

Blyth Road in wartime, closed by the Ministry of Supply. The Works fire engine and ambulance emerge from their garage at the front of the record factory on the right. Note the masked offside headlight and severely restricted other lights on the fire engine.

The construction of barriers at either end of Blyth Road further restricted entry to the site. It should be explained that EMI as a whole was undertaking a number of highly secret projects, not least of which were improvements to the radar equipment. This culminated in the development of a very sophisticated system compared to that available at the outbreak of war. Hence the requirement for such a high level of security.

Plans existed to relocate the factory as a whole or as individual units to Swindon in Wiltshire if enemy action had seriously damaged the production potential. A cottage was purchased in Uxbridge for a senior member of the management team to sleep in during air raids, in case they should be required in an emergency.

It was not only the staff who were affected, however, but also the product itself. Censorship of titles and their lyrics was undertaken to check for subversive content. Test copies [white labels] had to be produced for this purpose, prior to release for commercial production.

HMV arranged a display for the Ideal Home Exhibition at Olympia, London in March 1947. 78 rpm records were moulded, edge-trimmed and packed by a team including Laurie Richards, Frank Aldridge, Charlie Nash and Doris Dalziel-Buchanan.

A works vehicle decorated to take part in a 'V' for Victory Parade. Bald tyres were more fashionable in those days!

This operation was undertaken by a committee, including a Government representative, who had the final power of veto.

With the cessation of hostilities the factory, having successfully survived, once again resumed peacetime activities. Its next challenge, however, was far less dangerous, the introduction of the new microgroove records.

SPORTS AND SOCIAL ACTIVITIES

At the Blyth Road site the canteen formed the main social gathering point for staff, for many years. It was later replaced by a proper Sports and Social Club, in Printing House Lane, adjacent to the Blyth Road site. This club had excellent facilities, including both football and cricket pitches.

In the canteen, meetings were held for various clubs and organisations on most week day evenings. Enjoyable meetings of the Christian Union culminated, at the year's end, with a satisfying christmas dinner prepared by the canteen staff.

On another evening one could find a different club, being given a practical demonstration of a 'do it yourself' skill. Film shows were organised by the EMI film club, often giving members the opportunity to view films before they were generally released.

A flourishing Photographic Section gave many budding debutantes a firm footing in the art. Exhibitions were held which proved how competent some of them had become, even with the fairly basic equipment available to the amateur at that time. Another section that was to prove popular was the Gardening Club. This allowed members the opportunity to explore both exotic and practical ways of gardening.

Some of the comradeship associated with these canteen gatherings was lost when the Sports and Social Club was opened in Printing House Lane, in the mid 1960s. These were licensed premises, with ball room dances, and other entertainments, being organised regularly.

In 1972 the move to the new Uxbridge Road site was celebrated not only with an official opening ceremony, but also with an open day. This was held on the 10th September, and was for all members of staff and their families, so that they could view the new site in its entirety. This was to prove an extremely successful event, so much so that it was to provide a model for future celebrations!

This year also saw the beginning of the toolroom's domination of the Directors Cup for football. They won it not only in 1972, but also in the three following years.

In the summer of 1973, the record factory's cricket team, comprising Bill Johnson, Laurie Stewart, Len Giebeler, Ron Lane, Roy Matthews, Ron Rowe, Anthony Jamieson, Eddie Chilver, Cyril Gadbury and Eddie Wilcox, beat the team from Manchester Square by five runs. Cyril was credited with some fine catches, and the match was won with a good stumping by Laurie Stewart of the drawing office.

A later match that summer saw them victorious again, this time by twenty runs. The team changed slightly, however, and now

1968 saw the retirement of two departmental foremen, at which presentations were made by Divisional Director, J.B. Stevenson and General Manager, W.L. Rand. Above Harry Pickett, Toolroom and below Bob Lindley, Pressroom, with their staff, friends and colleagues.

The 1973 Record Factory Cricket Team also included Ken Cass and Ron Sarahs on the occasion of this photograph, which Eddie Wilcox missed.

After serving 46 years, Cliff Carter retired in September 1974. Pictured with Roy Matthews are Joan Adams, Alice Richardson, Bridget Lawrence, Joan Goodren, Tom Sayer, Brenda Hitchcock, Jean Harvey, Irene Smith, Alice Kennett, with Michael May and Don Eburne just visible.

included Alf Dewdney, John Tagg, Ginger Nichols, Ken Cass, Ron Sarahs, Bob Bailey and Roger Shenton, along with some members of the previous side.

A much more serious event occurred in 1974, when two men attempted to snatch the wages from the site. They were foiled by some of those for whom the money was originally intended. John Tagg, Edward Byrne, John Marshall, Eddie Chilver, Ron Sarahs, Alf Moody and Colin Mills [of ICS], were rewarded by the company for their bravery.

Among the other notable events in 1975 was a presentation to Jim Perryman, the Matrix and Label Foreman, who retired after completing 50 years service.

1976 was a year for some notable retirements from among the long serving staff. They included Jack Bates, Alf Ashworth and Jack Callander, all with 50 years in the company, Bill Johnson, with 48, and Nellie Ives with 39.

Nipper was assaulted during this year as well, by a lion cub! Brought in by a representative from Longleat, to promote a factory

Cyril Gadbury with long service staff – Jack Callender, Jack Bates and Bill Harris.

outing there, the cub was escorted on a tour of the factory. When the lion cub saw the plaster model of Nipper he seemed to take an instant dislike to it, and promptly "nipped" him!

The following year, 1977, was the fifth anniversary of the opening of Uxbridge Road, and the Queen's Silver Jubilee. In honour of both occasions an open day was held at the site. Over 8,600 people attended. Among the attractions, apart from seeing the factory itself, was a barge trip along the adjacent Grand Union Canal, run by John Ratcliffe, dressed as Long John Silver, with a parrot on his shoulder, accompanied by Nipper. An open top bus and a barge operated between the Blyth Road and Uxbridge Road sites, taking visitors back and forth. There was a mammoth balloon race, Len Giebeler in charge, and over 5,500 balloons were released. For the younger visitors Alf Dewdney performed magic tricks.

The site's first sports day was held in 1978, organised by Ted Riley. It attracted over 1300 attendees, with varying degrees of sporting prowess being displayed - particularly among managers, and their abilities to remain on board a donkey!

The open top bus outside Distribution reception area collects more passengers during the 1977 Open Day celebrations.

Len Giebeler with his Balloons! Margaret Butcher hides behind one, as Pam Giebeler looks on.

1979 saw the retirement of Dave Ridout, of the drawing office, after 43 years' service, and Dick Nichols, with 59 years! Ted Talmadge and Arthur Nicholas also retired, with 50 and 43 years respectively. Geoff Webb, who had for some years been involved with organising visits to the site, and had previously worked at a very senior level in the company with L. G. Wood, also retired.

Another very successful sports day, again organised by Ted Riley, took place in 1979. It was attended by even more members of staff and their families than the previous year. Events included "welly-throwing", a

On the occasion of his retirement, Dave Ridout accompanied by wife Peggy with friends and colleagues in the Engineering Office. From left to right, John Simmons, Colin Brown, Dave Bowring, John Pemberton, Chris Adams, Bob Kew, Nick Furtek, Geoff Pullen, Peter Williams, Roger Merrison, Les Zouch, John Byfield, Frank Bolger, Bill Soby, Alan Thompson, Len Giebeler and Ted Riley with Dog!

donkey derby, sack races and a tug of war, in which the prettiest team was undoubtedly the Record Factory Ladies team, led by Gloria Best.

The sites 10th anniversary celebration took place on Sunday 20th June, 1982. During the afternoon demonstrations of gramophone record and cassette manufacture occurred, and the methods of distribution explained. The open-top bus was again provided, for trips between Uxbridge Road and Blyth Road.

In the evening the events included a barbeque and childrens disco, held in the distribution loading bay. A caribbean atmosphere was created by the steel band in a marquee, and the evening concluded with a firework display. A special record was pressed to commemorate the occasion, and given to staff.

The upsurge in interest in running in the 1980s was reflected at EMI by the formation in 1984 of the EMI Joggers. At its height the club had 34 members. It was started by Bob Chatterton and John Salisbury, and included Graham Wilson and Jackie Hinge among its most enthusiastic members.

The sight of the members of the club setting out in the pouring rain for "fun" was a familiar one to staff at Uxbridge Road. We are assured, however, that it was not all hard work, and much enjoyment and many happy memories were generated during these events.

EMI Joggers outside Uxbridge Civic Centre to participate in the "Hillingdon Half Marathon".

Whilst they achieved some remarkable feats of running during their 8 years as a club, most notably in the Hillingdon half-marathon, their work for charity was outstanding. They raised, via sponsored runs, a total of almost £10,000 for Leukaemia research and local charities.

The year 1985 saw the opening of the record manufacturing and distributions sites' own sports and social club. This was housed in the one-storey building in the north-western corner of the site, which had previously been occupied by personnel department. The company donated some £20,000 to provide equipment for the facility, and Ted Riley gave the club its name - "The Dog and Trumpet", naturally! The artists Chas and Dave performed at the opening party.

The club was run, primarily, by a committee of staff at all levels in the company, with sub-committees for the various activities. All employees, past and present, were entitled to join. Sporting activities at the club included pool, darts, table tennis, golf, ladies keep-fit and football. Other functions organised here included quiz nights, dancing, discos and christmas parties for retired members of staff.

The next year, 1986, saw the end of an era when one of the factory's greatest characters, Ted Riley, retired, after 40 years service, without ever having had a day off sick!

The Uxbridge Road site's 15th anniversary occurred in 1987, and a fair was held to celebrate. Events included a concert by the Boys Brigade Band, fairground rides, a life-boat and canal trips. The fair was also showing off a fire engine that, it transpired, was actually on duty, and might have had to rush off any time!

This was to be the last celebration of its type, and was certainly a day to remember. It was attended by over 3,500 members of staff and their families. The whole event was opened by, who else, Cliff Richard, definitely the choice of all the staff.

This year also saw the retirement of Wally Rand, truly the end of an era for the worldwide manufacturing facilities of EMI.

On the 10th March 1988, Uxbridge Road participated in a successful attempt to gain entry to the Guinness Book of Records. This involved producing a 7" record in 47 minutes and 20 seconds from being handed a master tape. The event was broadcast live on Radio One's

Wally Rand at his retirement function with Bhaska Menon, Chairman EMI Music Worldwide.

Frank Maloney, Pressman, John Simmons and Eddie Kidd, with the 'Record Breaking' disc.

Steve Wright programme, from the Guinness World of Records at the Trocadero, Piccadilly in central London. The master tape was brought to Uxbridge Road by Eddie Kidd, on his motorbike, and was the Proclaimers single, "Make My Heart Fly".

This year also saw the return of the Directors Cup for football to the site, with the record production team winning. They also retained the cup in the subsequent year's competition.

The same year, 1988, saw the departure of John Simmons, after over thirty years with the company, and the return of Peter Hall.

There were several notable retirements among the senior staff during 1991, including, Colin Brown, Jim Hughes and Ron Turvey, all with over 40 years service.

With the greatly reduced numbers of staff now on site, resulting from the decline in demand for vinyl, the concept of combining and slimming down the size of individual departments was successfully extended to the football teams. The result was that they went from strength to strength. Combined manufacturing and distribution teams won the Directors Cup in both 1991 and 1992.

December 1991 saw the closure of the "Dog and Trumpet", as, with the decline in staff numbers, it was no longer a viable proposition.

Jim Hughes is surprised by the camera as he checks product quality!

The facility was to be used for outplacement counselling for staff made redundant, to help them find new jobs.

After the large-scale redundancy on the 12th June, 1992 when approximately 220 staff departed from manufacturing, a party was held. Those invited to attend included all employees, plus those recently made redundant. Attendance was extremely high, and the dancing went on until the early hours, in the nearby sports and social club of another part of THORN-EMI.

There have been a number of clubs and societies formed by EMI staff that have survived both the move to Uxbridge Road and the reduction in staffing levels. Most notable among these is the Bowls club. This was formed initially by a group of staff who transferred to the record factory when the machine factory closed, in 1955.

Known initially as the Sausage and Mash Society, the clubs members were famous for their parodies of managers and the like, and produced stage shows for supervisors. They introduced bowls shortly after joining the record factory, but maintained their tradition of exposing at their events the mistakes and misdeeds of management, in hilarious style.

The bowls was not taken terribly seriously, and was initially run almost exclusively by Ted Riley; joining was by his invitation only! He became more democratic later in life, and appointed a committee, with himself as "gold badge", George Shew as "silver badge", Dennis Bendall as "bronze" and Howard Parks, the Fire Officer, as "cowpat". One's position in the hierarchy at EMI bore only an inverse relation to one's influence when on the bowls green!

The annual presentation of bowls awards, home-made by Ted [and the entire skilled workforce at EMI!], was the highlight of the year. The items awarded were usually humorous, and not always welcomed on the mantelpiece!

Ted's last presentation was in the autumn of 1990, and he sadly died in 1991. It is a tribute to him that the club is still going strong

The second generation "Sausage & Mash" members pose with their trophies, shield, 'T' squares and Dog. Left to right standing: Colin Brown, Alan Thompson, John Ratcliffe, Frank Carter, John Simmons, Tom Cook, Frank Bolger, Ron Rowe, Bill Thorpe, Ted Riley; kneeling: Les Zouch, John Pemberton, Dick Jennings, Fred Tanner, Chris Adams, John Worley and 'Jenny' (centre).

even now with 35 members, many of whom are either retired or are ex-employees, and use the club as a means of maintaining contact with the company and their friends in it.

The other ongoing social activities that have survived this period of dramatic change include a number of annual departmental reunions. These cover a wide range of departments, including the toolroom, and there are a host of other get-togethers, nights out, christmas dinners, etc., among ex-staff as well as current employees.

The factory has always had a strong tradition of supporting local and national charities. During the last fifteen years of its life it raised over £45,000 for local children's homes, including:

    Hatton Grove, West Drayton

    Mulberry Parade, West Drayton

Bourne Lodge, South Ruislip
Charles Curren House, Ickenham
Standale Grove, Ruislip
Merrymans House, Hillingdon
Amherst Road, Ealing
Peter Pan Day Centre, Hayes

The fund-raising activities were organised by the Children's Fund Committee, headed by Mike Brooklyn and Vic Cusmans.

The staff were equally as generous for all appeals, including the various disaster funds, and television appeals.

STAFF SALES

A Staff Sales Shop has been available at both Blyth Road and Uxbridge Road sites for the retailing of discounted goods, most of which were produced by EMI or THORN-EMI. These included television sets, kitchen appliances, etc. Over the years arrangements have been made, allowing the sale of other companies' products in these shops.

Gramophone records, pre-recorded tapes and, in the latter years compact discs have, however, always been major sales items. Originally the staff sales shop was run by the parent company, but in the late 1980s it became an independent organisation, following a management buy-out.

ENTERTAINING THE WORKFORCE

Entertainment of the workforce has involved live performances, recorded repertoire from the EMI catalogue, and radio programmes relayed over the internal tannoy system.

The longest running transmission relayed by this medium was the BBC Light Programme's "Music While You Work". This had as its signature tune "Calling All Workers", composed by Eric Coats. The programme was first broadcast during the Second World War, and consisted of two half-hour programmes, one in the morning, and the other in the afternoon.

Another, and for some far more important use of the tannoy, was to relay the starting and finishing times of each shift. This was done by transmitting a 1,000 cycle tone. It was triggered manually from the time office, with the Greenwich time signal being used for a daily check. For personnel working in the noisy area of the power house, the signals were backed up by a powerful steam hooter!

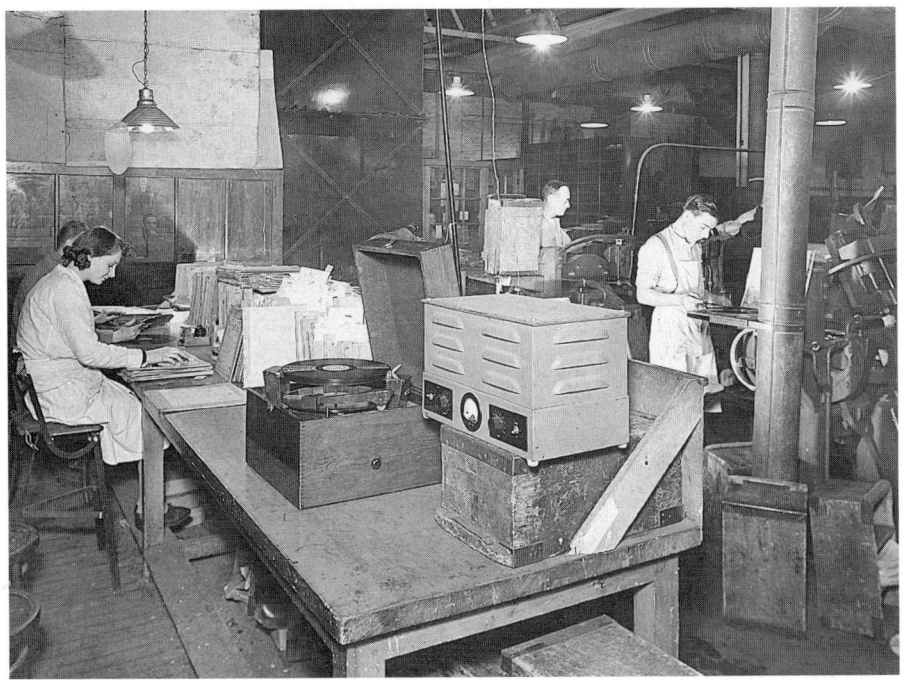

A record player with loaded autochange mechanism and amplifier unit to provide music for the pressroom and record examination department.

During the war the tannoy also served as an air raid warning system with pre-arranged codes giving the 'take cover' and 'all clear' signals.

There is a saying that there is nothing new under the sun. Why quote that here, you ask. Well, reading the publicity associated with the launch of an internal broadcasting station at Uxbridge Road, one would be forgiven for thinking the idea to be original. In fact the latest releases were relayed to the pressroom over the tannoy system in the 1930s.

The station launched at Uxbridge Road was named Radio E.R.B.S. [EMI Records Broadcasting Service], and was located in a converted tape cassette quality control room. It was, however, to be more professional than its predecessor. The station officially opened with a guest disc jockey in attendance, although it was normally staffed by three disc jockeys, who were more usually employed in the calibration department.

Initially it broadcast three times a week and soon became a firm favourite with the workforce who sent in many requests. So many, in

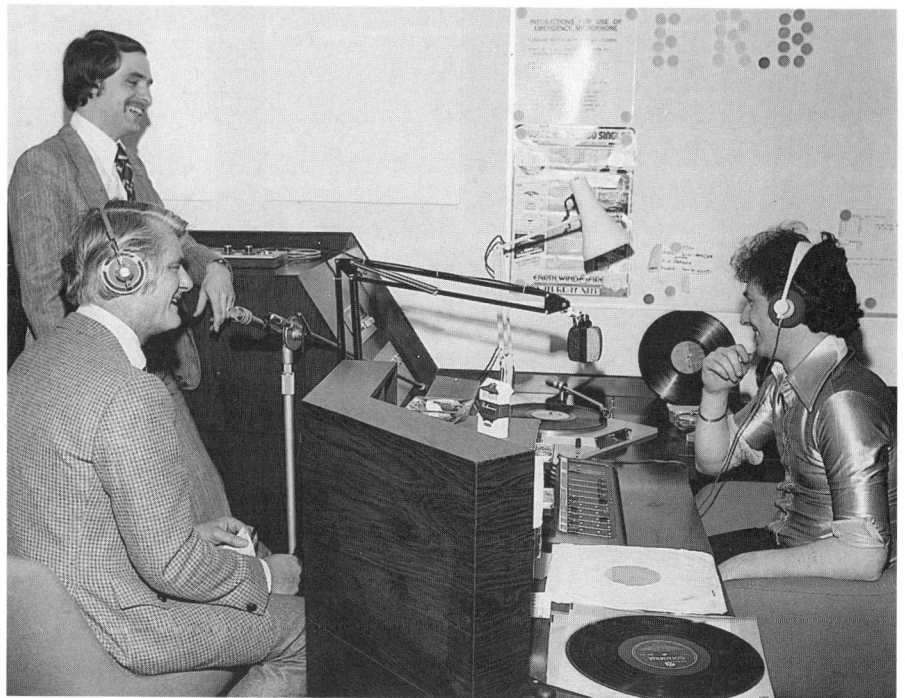

Pete Murray opens Radio E.R.B. with Malcolm Read at the controls watched by Peter Smith.

fact, that the original number of disc jockeys was increased, as were the broadcasting hours. A Revox tape machine was installed to allow the night shift to hear recordings of the broadcasts.

Christmas was a busy time for all the presenters, as requests poured in, and plans were finalised for all day broadcasts. These included games and competitions, in the run up to the Christmas break.

The expansion of the cassette department caused them to request the return of their quality control room. This led to the station being relocated into a converted matrix quality control listening room.

Despite the station's new home, the future of what was, after all, something of a luxury, became very cloudy. In order to save money the company closed the facility. A total shut down for only three months was the result, however, because of a resurgence of interest, involving deputations to the management from both the union and workforce.

The decision was reviewed, and the phoenix arose as Radio M.D.S. [Radio Manufacturing and Distribution Services]. It had new disc jockeys, and uprated programming under the overall control of Bill Rothwell. Staff reductions and pressure of work meant that the station would never regain its cult status.

Control of the radio station eventually passed to the Communications Manager, Mike Brooklyn. The number of disc jockeys had steadily been eroded by redundancies and reorganisations. In its last days the station became a background music medium, playing compact discs on a multiple disc player.

Radio M.D.S. finally closed down on the 8th May, 1992, a victim of the decline in demand for vinyl, and the resulting reduction in the size of the workforce.

VISITORS TO THE SITES

The sites have been visited by a wide range of people over their lifetimes. These have included crowned heads of Britain as well as overseas royalty. Politicians have visited the sites as have local dignitaries.

Now to scotch a commonly held myth. The public think that personnel working in the record factory would meet, almost on a daily basis, stars associated with the company's recording labels. This is not the case, visits by recording stars to the production sites are rare. When they do come each is given a warm welcome by the workforce.

Live concerts were given by both popular and classical artists in the Blyth Road canteen, during the war, and many artists were to visit the site.

The Beatles visited Blyth Road, and the crowds who turned out to welcome them were far greater than for any other visitors, despite the visit being given deliberately low-key publicity.

The years that the manufacturing plant was at Uxbridge Road saw the growth of the cult of the "pop star", and

Not all the visitors were well known – Peter Hall explains the matrix process to a senior representative from BASF.

Uxbridge Road staff enjoyed many visits by the famous and nearly-famous of the day.

In 1972, at the official opening ceremony for the new site, several celebrities were present, including David Frost, Anita Harris and Vera Lynn.

Early in 1973, Rolf Harris gave a short concert for staff when he visited, and other notable visitors that year included Charlie Drake, Geoff Love, Richard Baker, George Mitchell, and Gary Glitter.

1975 saw a concert by Geoff Love and Mrs. Mills, in the low bay area of the cover store, which proved immensely popular, and not just with the staff. The artists were so impressed at the sing-along abilities of the staff, that they considered recording an album there! Other artists to visit that year included Peter Noone, Don Estelle and Windsor Davies, Dave Clarke, Ken Dodd and Olivia Newton-John.

Visitors to the site in 1977 included Cliff Richard and Billie Jo Spears, and there was a host of visitors the following year, including Terry Wogan, The Kings Singers, and Cliff Richard.

The Wurzels gave a concert, in 1978, as did the Central Band of the Royal Air Force, and John Dunn of BBC Radio 2 visited the site to

Olivia Newton-John with Roger Black, Vic Cusmans, Pat Fay, Harry Donaldson, Jack Jeffreys, Mat Harry, John Campbell, Dave Scully, Rod Marlow, Joe Blackmoor; Mick Murphy and Bert Butler are in the background.

Percy Hemsworth tries Ken Dodd's hat for size as his mates look on; they include Jack Cole, John Clark, Mick Murphy, Roger Black, Reg Stone, Vic Cusmans, Bob Wallbanks, Roy Kerr, Albert Montebello, Albert Essling, Bill Partington, Tony Keeler.

interview staff, in order to explain to his audience how a record was made. Other visitors included the Mayor and Mayoress of Hillingdon, and Dennis Waterman.

The following year saw visits to the site from, among others, Gilbert O'Sullivan, David Soul, Max Boyce, Hot Chocolate, Pam Ayres, Dr. Hook, Charles Aznavour, and Kate Bush. There were also two concerts, one given by Instant Sunshine, and the other by Geoff Love and Lorraine [Luton Airport] Chase, in the cover stores area. The factory also became involved with Jimmy Saville, and his "Fix It" programme, when two youngsters spent a day there. Geoff Webb took them round, and "Butch" Bushnell helped them to make their own records on the laboratory press.

In 1980, despite the problems caused by the reductions in staff numbers, there were still several visitors welcomed to the site, including Sheena Easton.

Seen here with Cliff on one of his early visits to Uxbridge Road are: left to right, Margaret Ward, Pat Munro, Pat Vidal, Kay Hazell, Kath Frowen, Rose Moynehan.

Smiling for the camera with Cliff Richard are staff of the tape cassette department: Doris Serle, Claire Gowing, Peggy Sprake, Jill Congerton, Ann Merrick, Carole Bradshaw, Sue Fry, Theresa Cole and Marie Reed.

Billie Jo Spears was encouraged to take a "short cut" through the drawing office and is seen here with Peter Watson, Martin Bishop, Nick Furtek, Colin Brown, Bob Kew and John Ratcliffe.

John Harris, Matrix Department Foreman, explains the process of replicating metal parts to Gilbert O'Sullivan, as Roy Matthews prepares to interject.

Pam Ayres 'launched' this prototype 1400 press designed to sleeve 12" records with high density polythene bags. Bemused onlookers include her agent, a Laboratory Technician, Roy Matthews and Robin Allen.

Dick Hares looks on as Roy Matthews shows Kate Bush how labels are printed.

Charles Aznavour with the ladies in the Music Test including: Irene Partington, Mabel Morris, Sandra Baggs, Rose Cready, Sally Fox, Shirley Clements, Jean Smith, Betty Brooklyn, Daphne Fever, Dot Percy, Bridget Kiely and Joyce O'Donnell.

Tom Sayers, John Yeoman, Mike Smith and Gloria Hughes pose with Kim Wilde.

"Little, Simmons and Large!!"

Visitors in the following year included Kim Wilde, and Cliff Richard. Cliff was probably the most frequent and welcome guest to site; his support and friendship for it was always deeply appreciated by the staff.

The year 1982 saw a visit from Colin Crompton, the comedian, who noted in the visitors' book "fab place, nice people". Others that year included Arthur English.

In 1983 the site was visited by Kajagoogoo, a popular band of the time, and the disc jockey Mike Smith of Radio 1, who gathered some material for his show from the tour of the site.

Sacha Distel sent the ladies' hearts fluttering in 1985, and other visitors that year included Village People, a popular American band.

The visitors to the site during 1986 included Su Pollard. Regrettably, however, visitors during the later part of the factory's life became more rare. They included Cliff Richard, of course, who returned to celebrate the pressing of his 100th single release. He was allowed to make his own record on the 7" automatic presses, with just a little help from the press operator!

The last visit to the site by Cliff was in 1992, when he came with two objectives in mind. The first, and the saddest, was to say goodbye to the staff in distribution, whose operation had been moved to Leamington, in Warwickshire. The other was to accept the BS5750/[ISO9002] certificate awarded to the factory by Lloyds Register Quality Assurance, for all the staff who had worked so hard to make it possible.

HEALTH AND SAFETY AT WORK

The workforce at Blyth Road had a comprehensive fire and medical facility to call upon. The parent company operated its own ambulance and fire engine service. Both were manned by competent staff who were employed on a shift basis, to give a twenty four hour cover.

The ambulance was backed up by a comprehensively equipped surgery with trained personnel, and a company doctor available if required. If the injury was too serious, the county ambulance and hospital services were called upon.

The worst and potentially most dangerous fire took place at Blyth Road in April 1958, when the lacquer coating plant exploded into flames. This lacquer was produced from a base component made from 98% nitrocellulose [gun cotton].

Luck was on EMI's side, however, as the eighty gallon storage tank had been filled earlier in the day. This meant that the fumes could not build enough pressure to rupture the tank. This accident could have halted record production for some time as the record pressroom was next door, with the material compounding department close by.

Another kind of health care was available within the personnel department at Blyth Road, namely the welfare department. They dealt with any personal problems affecting the staff, including finding living accommodation for people who had been transferred to Hayes, for training within a department or in the apprentice school.

Following the relocation of the disc factory to the Uxbridge Road site, a separate surgery facility was maintained, located at the entrance to Gate 2. It was later moved to the distribution site, adjacent to the canteen. It had trained nurses, who were supported by a team of First Aider's. These were staff who had received basic training in first aid, and could render assistance prior to the arrival of the surgery staff.

Safety officers have been employed in both sites. They are responsible for interpreting the complex safety legislation, and ensuring that the company complies with the requirements, as well as encouraging safe working practices. They are supported by a Safety Committee, who monitored working practices in order to minimise the number of accidents. The company received several awards from the British Safety Council for its record on safety in the workplace.

The fire officer, however, now had to call on the County Fire Service rather than the Company's fire engine, as this facility was not provided at Uxbridge Road.

Peter Kenny receiving a Safety Award from the British Safety Council.

Regular fire practices have been conducted on both sites with some very commendable evacuation times being achieved, especially on warm summer days!

TOTAL QUALITY MANAGEMENT, ACHIEVEMENT THROUGH PEOPLE

EMI has always had a loyal and committed workforce, whose ingenuity and loyalty have contributed to the company's success. This was evident not only during the difficult times of the 1930s depression and the war, but also in recent times.

An example of this occurred in 1988, when the manufacturing plant was in somewhat of a turmoil. A major new manufacturing contract had been taken on board, to counteract falling volumes, but the plant was running at a loss.

A new senior management team in the form of Managing Director, Jim Leftwich, and the then General Manager, Manufacturing, Peter Hall, devised a strategic business plan to quickly bring the plant back to viability and improve customer service. The plan was called

'Framework for the Future', and was in two parts. Part I was to change working practices to minimise the amount of product handling, the wastage of time and materials, and the use of labour. In so doing the aim was to improve efficiency and effectiveness, and reduce the lead times for product delivery to customers. Part II of the plan was to focus on customers needs, both externally and internally, and to devise ways of meeting them on a consistent basis. The method used was Total Quality Management [T.Q.M.], achieved through management commitment to the idea that everyone in the factory would be responsible for improving the way things were done, and the implementation of BS5750 Quality Management System.

Jim Leftwich.

Part I of Framework for the Future was achieved by ensuring that the changes necessary were generated with the aid of, and, eventually, 'owned' by the operating staff and the introduction of incentive bonus schemes were used to encourage participation. The key players, and their areas of responsibility in the success of Part I were:

Jim Leftwich	communication and negotiations
Peter Hall	project director
Nick Wilkins	newly appointed manufacturing systems expert
Mike Russell	pressroom and exam
Geoff Grimmel	music cassette
Ron Turvey	technical services
Robin Allen	matrix
Mike Smith	disc planning
Colin Brown	engineering

The dramatic increases in productivity are demonstrated in the bar charts, Figs. 1 and 2, and the plant returned to profitability in 1990.

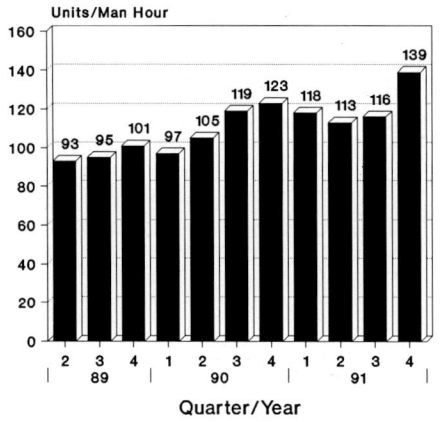

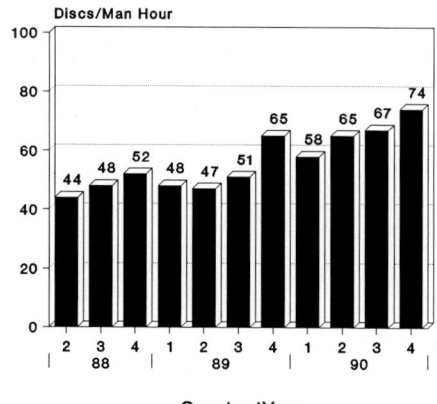

Figure 1.   *Figure 2.*

During the hard push of negotiating and implementing Part I, preparation for Part II was under way. A quality project team under the leadership of Colin Brown - Special Projects Manager, who had previous experience of managing nearly every factory department, began to research methods and companies already on the 'quality road'.

Colin's quality team comprised:

Jim Wilsher	Cassette Engineering Manager
Nick Wilkins	Manufacturing Systems Engineer
Jim Hughes	Chemist, and packaging expert
Robin Allen	Matrix Manager,
Leigh Pollinger	Purchasing Manager.

The first steps of T.Q.M. were to define the problems that existed and to do this a customer and workforce survey was conducted by specially trained staff. The results were disappointing to say the least, suggesting that only poor performance and negative feelings were experienced by the staff interviewed.

The management team now had to analyse the results and differentiate between the real problems and the perceived problems. Having a clear understanding now of the problems, the management team was then to create a vision of the future, to determine what the organisation would look like when the task was completed. Vision statements, devised and agreed by the team, were based on the concepts of:

- Customer requirements and satisfaction
- People and organisation
- Knowledge and information
- Processes and systems
- Continuous improvement
- Management and leadership

To aid the process, many experts were consulted and they included the Open University and John Oakland, Professor of Total Quality Management at the University of Bradford Management Centre. Two consultants in particular, Gunther Kruse and Peter Lightwood, made considerable contributions to the programme.

Simultaneously the initiative to implement a quality management system to the BS5750 standard was well on its way, aided by another consultant. BS5750 was considered to be one of the fundamental tools of T.Q.M., providing a system of defining and controlling all of the manufacturing processes and systems, detecting problems and rectifying those problems through Quality Improvement Groups.

The key concept of Total Quality Management is the involvement of 'everyone'. Part I of Framework for the Future had shown that the staff responded to being involved. T.Q.M. was designed to ensure the involvement of every member of the workforce.

The vision statements had identified the need for staff training to be undertaken in the principles of T.Q.M., and it was decided to train the entire workforce. The scale of the training was unprecedented. The training would be conducted by EMI staff, who would be trained to develop and deliver the programme by consultants over a 3 month period. A team of 6 volunteers were selected from across the site to form the training facilitators, they were Beverly Harris, Vic Cusmans, John Murray, Huw Saunders, Robin Allen and Martin Cook. Martin Cook, the Training Manager, was to leave the Company before the training began, but continued to provide guidance and advice.

The training material supplied by the consultants was considered by the facilitators to be unsuitable for the EMI culture, so they set about re-designing and restructuring the training package.

The final training schedule was completed early in 1991 and presented to the senior management. The schedule included 6 training modules each lasting approximately 3 hours, the 6 modules were to be phased over 4 months. The workforce would be split into groups of 12 to 16 members, each group being assigned their own facilitator and completing each module within the same time period.

The plan was accepted by the management, recognising the high cost and disruption to everyday business, and was seen as a demonstration of the Company's commitment.

It was also recognised that the administration requirements to get over 450 people to their training venues was formidable. An administrator in the person of Peggy Williams was assigned to the project. To minimise disruption, the group's training modules were scheduled to coincide with attenders' shift patterns - day, double-day or night shift - and each group was arranged to contain as wide a mix as possible of skills and status. This mixture of people also aided the participative nature of the training.

The training material, whilst still based on the consultants' text books, had been constructed in such a way that the workforce could relate to and understand the concepts of T.Q.M. in their own work environment.

To test this approach each module was previewed by Jim Leftwich, Peter Hall and a peer group, whose task was to critique the training material and approach.

By the end of February 1991, the T.Q.M. training had begun with module 1, which covered definitions of T.Q.M. and why EMI was doing it. The module also included the vision statements conceived by the management team at the start of the project. Module 2 was customer first, quality first, identifying the customer's needs and how to achieve them through the quality approach. Module 3, teamwork for T.Q.M., covered the advantages of working in a team. Module 4, putting T.Q.M. into practice used problem analysing and solving techniques for quality improvement. Module 5, people make quality, dealt with recognising that most problems are caused by bad systems, not by people. Lastly, module 6, recapped the whole course and tried to gain the commitment from each individual of what they could do to make T.Q.M. work. All modules contained a high proportion of discussion and participation, enhancing the 'everyone involved' theme.

The most important aspects of T.Q.M. are that it is management led, customer focused, depends on the commitment and involvement of everyone and on the process of continuous improvement.

Evidence that T.Q.M. was becoming effective was that by November 1991, 26 Quality Improvement Groups had been formed, involving over 150 people. Customer service was improving as demonstrated by the bar charts, Figs. 3 and 4, and the quality management system was approved to BS5750/ISO9002 standard by an independent assessment body, Lloyd's Register Quality Assurance in November 1991.

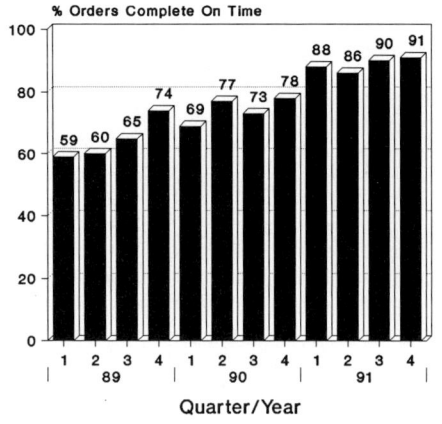
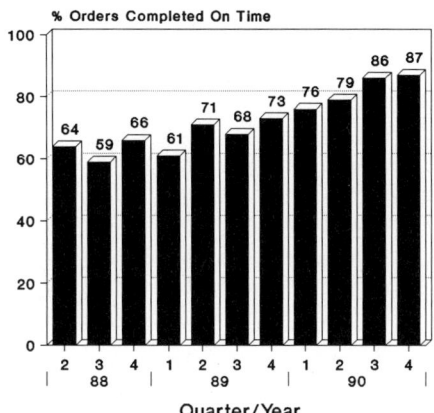

Figure 3.   *Figure 4.*

This was to be a great achievement. Hayes was the only vinyl disc factory ever to be approved to this standard, and the cost of the T.Q.M. training was recovered by the benefits derived from the Quality Improvement Groups within 6 months, truly a demonstration of the quality of the staff in the company.

AT THE HELM, MANUFACTURING MANAGERS

The first general manager of the record factory at Blyth Road was Mr. H. A. Irvine, whose appointment was announced at a Board meeting of the Gramophone and Typewriter Company Limited on 12th June, 1907.

Soon afterwards, Mr. Irvine visited the Hanover factory in Germany in order to gain an insight into the running of the factory, and the methods used there.

Quite how long Mr. Irvine remained general manager is not known, nor whether he was directly succeeded by Bernard Warner, who became general manager in 1935.

Bernard C. Warner	1935 - 1954
Harry D. Christmas	1954 - 1961
Walter L. Rand	1961 - 1969

Roy E. Matthews	1969 - 1978
John E. Simmons	1978 - 1988
Peter S. Hall	1988 -

MR. BERNARD C. WARNER

It was an exciting period of gramophone history when young Bernard Warner, then 24, joined "His Master's Voice" as a clerk in the record factory in 1912. Only two of the present large group of factories at Hayes were then in existence: the record factory itself, the foundation stone of which was laid by Nellie Melba in 1907, and the cabinet factory, where Tetrazzini laid the corner-stone in 1911.

In one capacity or another Bernard Warner figured largely throughout the fascinating story of progress and expansion of the EMI manufacturing facility. After he became Manager of the record factory in 1935, he played a leading part in establishing many significant developments, among them the technique of producing records from magnetic tape instead of waxes, and the many new processes involved in pressing microgroove [33.1/3 and 45 rpm] discs.

Under his expert guidance the record factory produced tens of thousands of matrices, not only for home use, but also to satisfy the ever increasing demands of the EMI factories overseas, while the output of finished discs ran into an almost unbelievable number of millions.

In over 40 years service to the industry, it was natural that Bernard Warner should have become a well-known figure; natural too, that with his profound knowledge and experience he should have earned the professional respect of his EMI colleagues and his innumerable friends in the gramophone trade, broadcasting and musical circles. More than that, however, Bernard Warner enjoyed the affection and personal esteem of all who were fortunate enough to know him and to work with him.

Bernard Warner died on the 15th December, 1954, aged 66. At his funeral, on December 20th, the Directors of EMI, his managerial colleagues, and representatives of the factory staffs gathered to pay their last tributes to a well-loved colleague.

Bernard Warner.

MR. HARRY D. CHRISTMAS

Born on the 2nd June, 1919, Harry Christmas joined the company in 1938, having previously been employed by a firm of consultant chemists. Continuing as a chemist in the works laboratory under the Chief Chemist W. E. Lord, Harry was to meet his future wife Zoe who was then Mr. Lord's secretary.

During the period 1939-45, Harry was engaged on work at EMI associated with the war effort.

Harry Christmas.

By the early 1950s, Harry was well involved in the preparations for the production of the long playing record, and was appointed Record Factory Manager in December, 1954, following the death of Mr. Warner. In 1957 he became a Director of EMI Records.

He was next appointed to succeed Mr. A. E. Newland on his retirement from the position of Manager, The Gramophone Co. Ltd., Dublin and Electric and Musical Industries [Ireland] Ltd., Waterford in 1961.

Harry continued in this role until appointed Managing Director of EMI [Industries] Pty. Ltd. and Director and General Manager of EMI [South Africa] Pty. Ltd., from the 1st February 1970.

Just over two years later Harry left the Company, but remained in South Africa establishing his own company specialising in manufacturing plastic mouldings.

Harry passed away in December 1990 at the age of seventy-one.

MR. WALTER L. RAND

Walter Leslie Rand, affectionately know to his many friends as Wally, joined the Company as a fifteen-year-old management trainee in 1934, and worked in the production engineering process laboratories. He studied mechanical and electrical engineering at Southall Technical College and, later, plastic technology at the Borough Polytechnic.

At that time he had no real intention of staying at EMI, as he wanted to become a pilot in the R.A.F., and although accepted as a

Wally Rand.

potential pilot, Wally recounts that "They sent me back after a few days, saying that my talents would be much better employed with the Company"!

In 1939, now qualified as a graduate of the Institute of Mechanical Engineering, he worked in the tool design office and, on the outbreak of the war, was appointed Production Engineer responsible for the manufacture of many million fuses. At the end of hostilities, Wally was responsible for dismantling the fuse production lines and the disposal of surplus equipment.

By 1947 he was engaged in development work associated with the long playing records, and the new magnetic tape in the works laboratories. He became a member of the Institute of Mechanical Engineering in 1948, and in 1950 rejoined the production engineering department then under the control of Mr. J. B. Stevenson. In 1956 Wally was appointed Engineering Manager of the record division. On Mr. H. D. Christmas' appointment to Managing Director of Ireland, in 1960, he was to become General Manager.

Mr. Rand established a strong engineering office, which became heavily involved with the modernisation of factory plant and the extensive changeover to the manufacture of LP records. Wally Rand assumed engineering responsibility for overseas factories in the early 1960s, and travelled extensively in the course of these duties. New factories were built in several locations, including Spain and Nigeria, and many others were updated.

In 1969 he relinquished responsibility for manufacturing, when he was appointed Manager Overseas Technical Services and Liaison, which was eventually located at Gloucester Place. New factories continued to be established and older plants closed by this team. Video disc production was established at EMI Electrola in Cologne and at a new plant in Swindon, Wiltshire. This was followed by the establishment of compact disc manufacturing at Swindon.

In 1980, Wally was made an honourary member of the Audio Engineering Society, in recognition of his many contributions to the art, and he finally retired in 1988 with the title Director, Production and Engineering Resources, having completed 54 years service.

MR. ROY E. MATTHEWS

Roy Edward Matthews joined the Company as an apprentice in 1952 with EMI Electronics. His first encounter with the record factory occurred when he transferred to the engineering drawing office in 1957. One of his first tasks was the design of the prototype vertical spindle plating unit, with five cells. After a period of development and trials, the mark II version evolved in 1960, and with few modifications, continues in use today.

Roy's association with the matrix department continued as he developed into the role of Process Engineer. He later undertook a comprehensive period of familiarisation with all production departments as part of his factory training.

In order to widen his experience, Roy left the Company in 1964 to join Fischer Price, a manufacturer of plastic toys.

He rejoined the Company in April 1968 as assistant to Wally Rand, and just one year later was appointed General Manager of the record factory. The success of the transfer of the manufacturing operation to the new Uxbridge Road site resulted in his appointment as a Director of The Gramophone Co. Ltd., in September 1972. Further promotion to Director, Manufacturing and Distribution Resources, of EMI Music Operations occurred, following his award of a Master of Arts Degree in Management Studies by Brunel University in May 1978.

Roy left he Company in September 1979 to fulfil a long held wish to set up in business of his own.

Roy Matthews in jocular mood at Don Woodward's retirement on the last day of 1975 after 50 years service.

MR. JOHN E. SIMMONS

John Edward Simmons joined EMI Electronics as a sixteen-year-old draughtsman apprentice in 1956. He transferred to the record factory as a Production Engineer in 1961, and worked on various design projects of record manufacturing equipment until 1966.

The next seven years of his career were devoted to the development of an injection-compression process for the manufacture of gramophone records, utilising injection moulding machines.

John initially worked closely with EMI Italiana Spa, who had undertaken much pioneering work utilising Metalmeccanica injection moulding machines. A new manufacturing facility was established in Beirut in the Lebanon with this equipment supplied from Hayes, and was John's first overseas project.

The Windsor SP6 machine superseded interest in the Metalmeccanica machines and led to the development of the SP130 version making it also suitable for 12" moulding.

The SP6 machines were installed in a new manufacturing facility opened in Bombay, while all later installations were to utilise the

John Simmons (left) with Roger Shenton (right) at a presentation to Ron Rowe on leaving the Company after 33 years service in 1980.

SP130. Principal among these other installations were the new Uxbridge Road facility, and a large installation at the RCA Records Indianapolis plant. This was to take John to America for the first of several visits in early 1972.

With the 7" injection compression presses settled into normal production, John was appointed Tape Records Manager in 1973, at a time when the 8-track cartridge was still popular. The music cassette continued, however, to find favour and soon captured this market. Ever increasing demands for this format resulted in EMI's decision to transfer cassette production to a new site at Winsford in Cheshire.

By 1976 plans were well advanced and it was ironic that on the very day that John and his family moved to their new house in the north, the project was aborted by the EMI Records Board of Directors!

Returning to Uxbridge Road, John was appointed Deputy Factory Manager with responsibilities for both disc and tape. Two years later in 1978 he became Manufacturing Manager. In the autumn of 1979 an ongoing inadequate workload resulted in the disbanding of the nightshift and the workforce was restructured into double day shift operations.

The resulting reduction in overheads led to the operation returning to profit once again, assisted considerably by the acquisition of a major manufacturing contract from RCA Records early in 1970. To cope with the increased demands and heavy workload, overtime working was frequently necessary, and a new enthusiasm was generated in the workforce.

An appointment to General Manager followed in 1981, and after the establishment of EMI Manufacturing & Distribution Services, John was appointed Director of Operations in 1984, assuming responsibility for both manufacturing and distribution. He was also Deputy Managing Director, under Ted Harris, and continued in this role until leaving the Company in 1988.

MR. PETER S. HALL

Peter Scott Hall joined the Company in 1977 as a twenty-seven year old manufacturing financial accountant, having gained previous experience in the employ of British Oxygen Company, H. J. Heinz and elsewhere.

Progressing to manufacturing management accountant, Peter then changed direction, with his appointment as Cassette Factory Manager in 1980. He continued in this role until leaving EMI Manufacturing and Distribution Services in May, 1981, to gain further experience with another company in the same business.

Peter Hall.

As Director, Forward Sound and Vision, he was responsible for the Orlake disc manufacturing operation, notable for successfully capturing the then new picture disc market.

In 1988 Peter was encouraged to return to EMI Manufacturing and Distribution Services as General Manager, Disc Manufacturing. His responsibilities were soon further extended by his appointment as General Manager of Manufacturing.

With Jim Leftwich, Managing Director, Peter was co-author of a strategic business plan formulated to radically improve both the customer service and efficiency of the manufacturing operation. Entitled "Framework for the Future", it was launched in 1989, and was the start of a programme of improvements which ultimately lead to the manufacturing division being accredited to BS5750/ISO9002, in the autumn 1991. At this time, the only music company in the world to achieve such accreditation.

The closure of the CBS, Aylesbury, disc manufacturing facilities in 1990 fulfilled the ambition, contained in "Framework for the Future" for EMI to be the last major company manufacturing vinyl discs in the U.K.

1990 was a significant year for Peter, as he also became Director of Manufacturing for EMI Music Services, and was awarded his Bachelor of Arts Degree in Business Studies [1st Class] at East London Polytechnic.

Peter has proved popular with the workforce at all levels and maintains a high profile, being seen frequently in all production areas, accessible and approachable to all.

Peter continues to have responsibility for the operations of the manufacturing facility at Hayes.

CHAPTER FIVE

Anecdotes

This section contains some the more printable tales related to us during the course of compiling the book. They seemed not to fit into any of the other chapters, but seemed to good to miss out. They give some flavour of the more humorous aspects of working at the factories, it was not all work!

AN ENCOUNTER WITH THE BOSS

Bill Goodhew had just returned from holiday and was operating the direct inject presses, just after their commissioning. Things were not going well with the new machines. Of the eight then installed only one was working without giving constant trouble, and this was one of Bill's. It did, however, have a fault. The problem was, that although records were being produced, the hydraulics were playing up, causing extremely loud bangs every time it cycled.

Richard Burkett [Head of Operations Worldwide] who was visiting the site that day, complained to the supervisor about the noise from this machine. Bill did not know Richard, and when the supervisor told Bill to get the press stopped and the noise fixed he replied "It's the only press that's working, tell whoever he is to leave it alone [expletives deleted]!"

At that moment Richard appeared from behind the press and introduced himself!

GHOSTS IN THE MACHINE

It is not generally well known, but a cutting head can act as a microphone.

At Blyth Road, lacquers were cut for test purposes. Many were cut with no recorded sound, so that the quality of the surface could be assessed.

On this occasion the cutting operator, however, spoke quietly [so as not to produce any modulation marks in the cut, that would give the game away] near the recording head while doing the trial cut. He repeated over and over again "are you awake?"

The results were quite spectacular. The quality assessor, hearing apparently mysterious "voices", tore out of the room as though terrified. Dire results were promised by management for any recurrence of the event!

A GHOST IN THE BUILDING

With a factory having existed on the site since 1908 it hardly seems surprising that a ghost or two might haunt it.

After most of the factory had moved to Uxbridge Road, Chris Pooley was left on his own in the laboratory at Blyth Road. Twice he and his boss heard voices conversing in a foreign tongue in an adjacent room. The first time the room was locked but on the second it was open.

They rushed in and found themselves standing in a cocoon of warmth, both agreeing that whatever it was it was seemed friendly. The presence did take a dislike to one member of staff, however, who came down one evening to deliver some urgent work.

He ended up outside, covered in perspiration, despite it being one of the coldest nights of the winter. Needless to say, he never ventured inside the building again!

THE NEW GENERATOR

The whole day, at Blyth Road, had been punctuated by loud thumps from the direction of the power house. This was caused by a group of men trying dig a hole for a new generator. Despite using a ball they seemed to have been unable to break through the concrete.

They had, by 5 o'clock, decided to use explosives, and evacuated the power house.

The explosion rocked the floor and sent slivers of concrete through windows and switchgear, blacking out the site and cutting off power. Unknown to them, the concrete had actually broken up, but the ball had recompacted it!

THE NEW OIL TANK

To install a new 3,000 gallon hydraulic oil return tank in the south-east corner of the pressroom, it was necessary to dig a hole 10 feet by 10 feet by 16 feet deep. Unfortunately, at 10 feet deep the Yeading brook intervened, and the last 6 feet were dug underwater, with the assistance of a pump!

When the hole was finally completed it was discovered how tight the specification was. Ron Turvey, Ralph Laurent, Bob Lord and Bob Bailey had to practice synchronised jumping on top of the tank to get it into the hole!

THE DANGERS OF WELDING

Brian Spoard's car needed some work doing on it. He asked Tommy Hall to do it, one Saturday. Tommy was busy, but said he would get Bill Timberlake to do it. Bill, reluctantly, agreed.

There was one small problem, however. Unknown to Bill, the car's normal petrol pipe had been replaced with plastic tubing. The car actually caught light twice during the welding. The first fire was not serious, the second, however, was the end of the car.

It burned despite the efforts of several people to bring it under control. It was so hot the tarmac underneath melted, and the car wheels sank deep into the car park surface!

On the Monday morning an investigation was held and the culprits identified. They were all spoken to by their managers. Everyone waited for the retribution that must, surely, come from on high.

Nothing happened, however, and no-one was disciplined. Bill's comment, when told by his manager not to weld any more cars, bears repeating. He said "Funnily enough, no-one seems to want me to weld their car at the moment!"

THE TOW BAR

Eddie Walsh decided to buy a tow bar for the back of his car, and eventually got one, second-hand from George Hayden. The following morning he sought out George and told him that he couldn't connect anything to the tow bar.

George couldn't understand this, and they proceeded to inspect Eddie's installation. When they got to the car the problem became clear. Eddie had installed it upside down, with the hook pointing to the floor!

George explained this, but Eddie retorted, "Well, it was that way up when you gave it to me!"

FIRE IN THE PRESSROOM

Bob Bailey has a reputation for being unflappable. One Saturday he and several others were replacing the hydraulic oil valves, because they were underrated for the pressure they were carrying. This involved cutting them off with a torch, and fitting new ones. Bob Lord began cutting, without having the oil drained. The resulting 20 foot jet of flame set light to the lagging on the pipes in the trenches.

Eventually someone contacted Bob's manager, Ron Turvey, and told him the pressroom was alight. Ron duly arrived, in a high state of panic, and met Bob.

"Is there a problem," asked Ron?

"No," said Bob.

"I heard you had a fire," said Ron.

"Yes," replied Bob.

"What's the situation then?"

"Well, there was no need for you to come in. We've been fighting the fire with hoses all morning, and it will soon be under control! The job will be finished on time."

TRAINING ON THE NIGHT SHIFT

One evening in 1975, Roy Matthews [then General Manager] was walking through the pressroom, when Lofty stopped him and asked if he would have any objections to the nightshift doing some exercises, when work was slack. Roy had none, and said he thought it would be a good idea.

Picture the scene a few nights later, when Digger [Adrian Bigby], the Night Shift Foreman, went to check on production in the pressroom. To his amazement he found Keith Harris in the aisle doing one arm press ups, Lofty suspended from the "flying carpet", doing pull-ups, and Nigel Brooks also doing pull-ups, from the hook on the press hoist.

Three other operators were being shown how to do push ups, between their press and the bench, and two others doing one and two legged squats.

Jim Woods and Peter Dell were practising their running. They were doing two laps of the production site, checking their machines, then going back out for the next two laps.

Digger marched down the pressroom, heading straight for Lofty, who, by this time, had dismounted from the "flying carpet".

On reaching him, Digger demanded to know what was going on. Lofty replied, "It's okay, I asked Mr. Matthews yesterday if we could do some exercises, and he said he had no objections."

This threw Digger completely. He could not believe Roy Matthews had actually agreed with Lofty, and yet he knew Lofty well enough to know he wasn't pulling a fast one.

The next day Digger attended a meeting at which Roy Matthews was present. At the close Roy asked for any other business, and Digger asked him if he had agreed that the night shift operators could do exercises in the pressroom.

"Yes," said Roy, "I have no objections. I thought the handwriting practice would do them good, they don't get much of a chance!"

SMOKING IN THE PRESSROOM

Fred was on of the most laid back people in the company, with a reputation for calmness that was legendary.

On this occasion he and Daphne were going round the press shop, together, collecting up records for testing. Daphne asked Fred if he could smell something. Fred said no. She persisted that she could smell burning. Fred couldn't, and was beginning to show signs of irritation.

Daphne, however, would not be put off, and persevered. Fred was now definitely irritated. When Daphne turned to look at him she discovered exactly what it was she had been able to smell. It was Fred that was alight!

Fred had been slowly smouldering from the waist upwards, and the front of his jumper had a big hole burnt right through it. This was getting bigger by the minute, as was the one in his shirt! It transpired that his hand-rolled cigarette had fallen from the corner of his mouth, and slipped down between his jumper and shirt!

By this time Daphne was slapping Fred's jumper frantically, in a desperate bid to put him out, which was eventually successful.

Fred, however, was unconcerned about the danger to himself. He was more baffled about the cause, and seemed only just to be able to tolerate Daphne putting out the fire. All he could say was "I wondered where that roll-up went, I thought they were supposed to go out on their own!"

HELP FROM THE MANAGER

One Saturday afternoon, several urgent orders were running through the plant, and the flash reclaim system had broken down. The flash was everywhere. Operators were helping to keep the pressroom clear, but there were bins and cardboard boxes of the stuff everywhere.

The supervisor, Harry Fox, however, was not helping to move the flash. "Not my job" was his view. The telephone rang. It was Mike Russell, the manager. "I'm coming in to help," he said.

The supervisor panicked, and began helping clear the flash, getting filthy dirty like everyone else.

Mike arrived to "help" clear up this problem. He was wearing white Gucci shoes and trousers, with a pale pink jumper and socks. Needless to say, he left in the same degree of sartorial elegance that he arrived, not having lifted a finger. The supervisor, however, was by now covered from head to toe in dirt!

What he is alleged to have remarked after Mike's departure is unprintable!

FAITH IN EXPERTS

A revolutionary new LP inner bag was developed by Bill Soby and Paddy Marchant. The bag was a blow moulded tube, cut and welded to form a 12 inch inner bag. Many technical problems, not least dimensional stability, were encountered.

Paddy was undaunted by the vast number of problems that arose with this bag, and to each he would respond, "I know what the problem is, and I'll get it solved for the next batch!"

Those involved in trying to run these bags were beginning to get a little fed up with the project, and Paddy's responses. They decided to set him a test. A dozen bags were obtained, and they guillotined one inch off the top of each.

Sure enough, when presented with these bags Paddy responded, with the air of an expert, " Oh, yes, I know what the problem is, and I'll get it solved for the next batch!"

FIRE IN THE MATERIAL MIXING

A welding job was being done on the third floor of the material mixing department. A little while after it was noted that the hopper containing the additive Lectro 78 [tetra basic lead fumarate] was getting hot.

Bill Soby, Laboratory Manager, was called. No problem, he said, Lectro 78 would not support combustion.

The hopper, however, got hotter. Bill still insisted it would not burn.

Ralph Laurent finally got a scoop, and opened the hopper to remove a sample of the material. The sample was glowing red-hot, and the heat flared up in the hopper as the lid was lifted. The fire brigade arrived summoned by Harry Morphew. They arrived in their dozens; no-one had ever seen so many fire appliances in one place before.

While the staff retired to sample Dennis Bendall's "Scottish water", the fire brigade dealt with the problem.

The material was emptied from the hopper, and eventually put out. Bill's final words? "I may have been not totally correct in my statement that it would not support combustion!"

THE PERILS OF SMOKING

Harry Fox was smoking one of his favourite cigars, a King Edward. He was called across to help Jim Mullally, who was having some difficulty with his press. Harry attempted to effect repairs, but needed to get something from his office. He put his cigar down on the bottom mould and left.

Jim returned, assumed his press was fixed, and started it up. As he did so, Harry returned, and immediately began looking for his cigar. When the press was opened, all was revealed. It was moulded into a 12" record!

TRANSPORT PROBLEMS

Paddy Marchant, the Laboratory Technician responsible for packaging, became the proud owner of a 50cc Honda moped, when they were first introduced.

Impressed by the economic reliability of the bike, Paddy tended to go on and on about it. Bored to breaking point, Robin Allen devised revenge.

A puddle of dirty motor oil was strategically placed under the bike. Paddy was astonished and proceeded to investigate. Puddles of oil continued to appear for many days much to Paddy's horror. After a while the tampering ceased. When a couple of days had passed without oil leaks, Paddy announced that he had solved the problem; "It was merely a faulty valve cover gasket". He looked bemused at the hysterical laughter that followed!

INDUSTRIAL RELATIONS

The toolroom convenor, Jack Collison, particularly wanted to see Roy Matthews one night, although he had no appointment. When he got to Roy's office the door was shut. He refused to believe that Roy was not there, so he sat outside and waited.

Roy did not want to see Jack, as he had another engagement, which meant leaving the factory earlier than usual. He was, however, effectively trapped in his office.

He soon found a solution to his problem, and, by telephone, arranged for a ladder to be delivered from the maintenance department.

He proceeded to climb through the first floor window, on to the ladder, and made good his getaway. It is not known how long Jack remained waiting for him!

THE UNION MEETING

Eddy White, then Personnel Officer, and Eddie Walsh, the shop steward for the pressroom, were having some difficulties. Eddie Walsh was pressing Eddy White for a meeting he really did not want to have.

In the end Eddy White grew so fed up with Eddie asking for a meeting that he agreed to have one, but would not say where. After much further argument he agreed to a venue, "We'll hold it in the canal," he said, "preferably with you at the bottom!"

COST CUTTING

During one of the many cost-saving initiatives that have taken place from time to time, John Simmons decided to set an example by economising on the cost of accommodation. This centred on a visit by himself and three others to Winsford. To show the savings that could be made, he booked the accommodation himself.

On arrival the lodgings proved to be inhabited by a road repair gang! Undeterred, the party set forth for an evening out. On return to the lodgings at about 11 o'clock, they discovered that the door was locked. After knocking on the door they were confronted by the angry owner leaning out of the window, threatening to refuse them entry. When they were eventually admitted, they were faced by the owner and his two, very aggressive alsatian dogs. He proceeded to give them a headmasterly ticking-off!

John not only never used the place again, he allowed engineers to book their own accommodation in future!

COMPUTERS

As part of the changes contained in Framework for the Future, an incentive scheme was introduced into the pressroom. This involved recording large amounts of detail concerning hours worked, production output, etc. Mike Russell, the Manager, soon produced an enormous computer spreadsheet to calculate the amounts payable.

Given the natural inclination of managers to delegate anything remotely resembling hard work, he promptly gave the job of using this programme to Stan Poole, the Pressroom Controller.

Stan was not entirely at ease with computers, and one day a cry of anguish was heard from his direction. He seemed to have lost the entire programme. The exact cause of this was never discovered, although the situation was recovered by Mike. This experience, however, just made Stan more worried and nervous.

A month later, Mike Russell and Robin Allen, seeking to exact revenge on Stan for various reasons, devised a scheme to make the disappearing programme problem recur. They copied all of the data onto a floppy disc, and then wiped the programme clean of all data.

Stan entered the office, switched on the computer, selected his programme and lo! A blank screen! His face turned a cross between white and green, and he began to stutter. Mike and Robin pretended to try to help him, but to no apparent avail.

The conclusion reached was that Stan would just have to re-enter the data, a task that would take several days. Had it not been for Mike's inability to maintain a straight face, Stan may well have been there today, still trying to input the data and work out what had happened!

QUALITY CONTROL

Jim Mullally managed to start his press without any stampers in it, inspect the product and carry on! He eventually made three boxes [165 discs] before the other operators could persuade him that there was no music on the records!

MEETING THE STARS

Su Pollard, star of "Hi de Hi", was visiting the cassette department. After going through the "Hi de Hi, Ho de Ho" routine with the staff she exclaimed, "It ain't 'arf hot in here".

Jill Congerton asked if she would like a cup of tea.

"I'd love one" replied Su.

Jill supplied the tea in her own cup, which disappeared never to be seen in the factory again.

The cup, however, is claimed by Jill to be the one Su was drinking from in an episode shown sometime later!

ROYAL VISIT

During a visit from HRH Prince Phillip, Wally Rand had organised the factory to unsurpassed levels of cleanliness, and he was especially immaculately turned out. Striding side by side with HRH through the factory, Wally was in his element.

Unfortunately, travelling the opposite way to Wally and HRH was a fitter known as Dominic [Herman Baker]. He was in charge of greasing and oiling, and even by "grease monkey" standards was particularly filthy, being covered in both oil and grease with the ever-present "dew drop" in attendance, along with his oil can and newspaper.

Wally was horrified, but could not avoid meeting Dominic [with whom he was on generally good terms] as there was no alternative route. Dominic sauntered past, and said "Mornin' Wally, wet old weather ain't it?" without recognising who Wally was guiding round.

It is rumoured that Dominic subsequently spent several days hiding around the plant, in fear of Wally's wrath, after someone had told him the full facts!

SUPERSONIC FLIGHT

There was a dispute on the shop floor, and a request was made for an urgent meeting with Wally Rand. The shop stewards were told, however, that Wally could not meet them as he was in the U.S.A.

Things came to a head, and after a lunchtime meeting the workforce decided to hold a stoppage unless talks were immediately forthcoming. Wally appeared as if by magic within the hour!

MOVING THE UNION OFFICE

The union office was sited in the distribution building at the time, and John Simmons wanted it moved, to enable the World Records Club warehouse to be built. Ron Turvey was given the task of moving it. He turned up on the Friday with a JCB and driver. John Murray, the convenor, was not amused, especially when he had an altercation with the driver over the cuckoo clock on the wall of the office and the likely effect on it of a JCB!

But, sure enough, the following Monday it was gone, and John found himself in the toolroom, west end, office. The room there was quite pleasant, and John was not too upset at the prospect.

John Simmons, however, was. John Murray was now occupying the room he meant to use for the laboratory microscope. Ron Turvey was again contacted, and told to move the union office to the adjacent room.

Ron rang the contractors, and told them to take the office down that Monday, before start of work. When Ron arrived on the Monday the contractor was waiting for him, and said he couldn't take the partition down, because the wiring was still inside it.

Ron, realising he would be in trouble with John Simmons if the union office was still in situ, contacted the electricians and asked them to de-wire the room. Unusually, for the electricians, they responded immediately, mainly because they knew how much of a stink would be raised if the job was actually done!

The office came down. John Murray was furious, and complained to both John Simmons and John Munro, but to no avail. Later that morning Ron went over to see John Murray, to discuss the building of his new office.

As he arrived John was walking down the fire escape. Upon seeing Ron he threw his kettle [a useless article, considering Ron had taken away both his electrical supply and his water tap] on to the floor and jumped on it in rage from the fourth step up! John was speechless, other than some unprintable remarks, a rare event indeed!

THE NEW SUPERVISOR

The foreman, shift controller and several other members of the pressroom team were sitting in the office having a break. Stan was reading in the paper about the "troubles" in Famagusta. The paper did not, however, mention where Famagusta was.

He looked up from his paper, and asked the foreman and the others where it was, but none of them knew. The foreman said that he would ask their new, bright and very keen supervisor, Peter Merton, who was bound to know. Peter was duly summoned to the office, and appeared, in his immaculate white coat, sharpened pencils in breast pocket, and notebook ready.

"Peter, where's Famagusta?", demanded the foreman.

"I don't know" came the reply, "but if you can tell me his clock number, I'll go and find him!"

THE ONE-WAY SYSTEM

The Uxbridge Road site has always had a one-way road system, and this diverts all traffic past the front offices, where the managers reside.

One Saturday Tommy Blair, who was known to like an occasional pint, had disappeared to a local watering hole for an extended lunch. While he was away, however, his manager, Ron Turvey arrived on site, and was looking for him. An urgent phone call from his colleagues caused him to rush back to the site.

Realising that he would have to pass Ron's office, and that Ron was bound to see him return, he entered the site, and went the wrong way round the one-way system. Tom believed he could then pretend to have been working in some obscure part of the site, where he could not be contacted.

Unfortunately for him, Ron had a fairly good idea of where Tom was, and what he would do. As Tom arrived at the department, Ron appeared from his hiding place in the cover store loading bay!

On another occasion, Mike Tucker cycled round the site, and was hit by a fork lift truck. The bike was a write-off. The Safety Officer, Harry Morphew, was having a campaign based upon the number of near misses that occurred. He told Mike to bring the bike in, because he wanted to use it in a safety display, and he'd make sure that he was compensated.

Mike went to see Ron Turvey, to get the money, but close questioning by Ron revealed the truth. Ron was not going to pay for damage caused by Mike going the wrong way round the one-way system. Mike was furious, because he'd told Harry the truth, and having been promised payment by Harry, took his case further. He was, however, never paid!

THE TORTOISE

Reg Gizzi bought an empty tortoise shell to work, and went into the examination department. He explained that he had left it there yesterday, and the tortoise had escaped from its shell. Could they please help him to look for it?

It was only the intervention of the supervisor that prevented the rest of the day being spent by the examination ladies in the pursuit of the animal!

MAINTENANCE WORK

Derek Fry and two others attended to the effluent pump, and Derek volunteered to go down into the pit as he had the wellington boots. Who flushed the toilet when he was in there, and filled his boots?

Big Mac [Macveen Underwood] held Tommy Blair over the parapet of the bridge between manufacturing and distribution by his ankles, as revenge for some misdeed. He kept him suspended thus for several minutes!

Ralph Laurent became the proud owner of an extremely expensive, brand new Meerschaum pipe. It had been impregnated with honey, to give the tobacco a special flavour. He bought it into work, proudly displaying this present from his family. One hour later he dropped it into the hydraulic oil tank, never to be recovered!

Tommy Hall had an accident with a pair of stillsons, and smashed all his teeth. He approached Ron Turvey and asked if the company would pay for the dental treatment. Ron agreed.

Six months later he presented Ron with a bill for £120. Ron was horrified "I shouldn't have to pay all of this, after all, these new teeth are better than the ones you had before!"

"Yes", Tommy replied, "but I couldn't take those out!"

THE NEW BOY [OR BE CAREFUL AT INTERVIEWS]

When Robin Allen came for his first interview at EMI, he was seen by Jimmy Hughes, who turned him down. Some weeks later he was interviewed by someone else, and was successful.

Guess whose boss Robin Allen ended up as, some 29 years later?

THE ROLL OF HONOUR

The Company has always recognised 'long service' for personnel, and circa.1970 additional recognition took the form of framed gold discs. The design on the disc depicted the length of service achieved. All the framed long service award discs were presented complete with the famous "His Master's Voice - Dog & Trumpet" label bearing the recipients name and commemorative inscription.

50 YEAR PLUS SERVICE

Few have come close to achieving 50 years of service, but as time goes on it will become even more and more difficult to reach this milestone with the Company. 1990 saw the 50 year award presented to Ron Crawley, consisting of:

The unique 30cm gold disc, etched with an aerial photograph of the Blyth Road complex as it was 50 years ago, together with facsimile signatures of the inventor of the lateral cut disc record, Emile Berliner, as well as three 'Golden Age Singers' from the early years of the gramophone, Enrico Caruso, Dame Nellie Melba and Feodor Chaliapin,

together with

A limited second edition model 'His Masters Voice' commemorating Nipper 1884 - 1984. Commissioned by HMV and produced in conjunction with Garrard, The Crown Jewellers, and artist Neil Campbell, by Albany of England.

40 YEAR PLUS SERVICE

A 30cm gold plated disc etched with 17 replicas of the Company's first Trade Mark 'The Recording Angel', as used in various parts of the world in the early years of the twentieth century, was the design of the 40 year award.

30 YEAR PLUS SERVICE

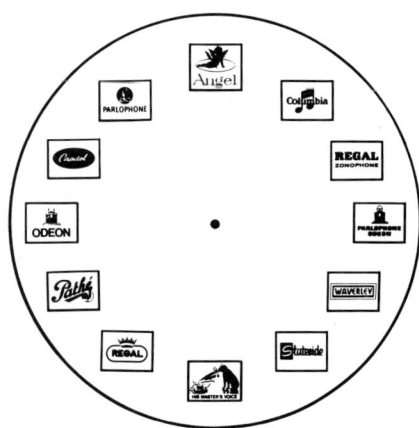

The 30 year award was a 30cm gold plated disc etched with twelve different Company recording labels.

20 YEAR PLUS SERVICE

The 20 year award took the form of a 30cm gold plated positive.

The following lists are of the recipients of the above awards, who completed their service at Uxbridge Road, in alphabetical order. Nearly all recipients completed more than the number of years shown on the awards, and they are traditionally referred to as the number of years "plus" awards.

Such loyalty to the Company is outstanding, and its gratitude to such staff cannot be over-emphasised. These staff formed the backbone of the manufacturing division, and made it the tremendous place it was.

50 YEARS PLUS SERVICE

Mr. A. Ashworth
Mr. J. Bates
Mr. J. Callander
Mr. R.A. Crawley
Mr. L. Morton
Mr. R. Nichols
Mr. J. Perryman M.B.E.
Mr. W.L. Rand
Mr. W. Soby
Mr. E. Talmadge
Mr. J. A. Wheeler

40 YEARS PLUS SERVICE

Mr. E.G. Bass
Mr. N. Bates
Mr. W.H. Birdsey
Mr. C.J. Brown
Mr. D. Browning
Mr. W.E. Bushnell
Mr. A.C. Carter
Mrs. J. Cummins
Mr. C. Gadbury
Mr. R.J. Gooch
Mr. E.D. Guley
Mr. F. Hanwell
Mr. W. Harris
Mr. J. Hughes
Mr. G. Jenkins
Mr. W. Johnson
Mr. J.A. Kennedy
Mr. P.J. Kenny
Mr. L.G. Kille
Mr. A.H. King
Mr. F.J. Leatherby
Mr. J.F. Leppard
Mr. R. McBride
Mr. J.E. Mooney
Mr. A.F. Neville
Mr. A. Nicholas
Mr. K. Owen
Mr. J. Pemberton
Mr. G.C. Price
Mr. D. Ridout
Mr. E.G.T. Riley
Mr. T. Sayer
Miss. K. Slater
Mr. F.W. Tanner
Mr. T.A. Thomas
Mr. W.O. Thorpe
Mr. R.J. Turvey
Mr. A.A. Warren
Mr. J. Yeoman

30 YEARS PLUS SERVICE

Mr. J.C.W. Adams
Mr. C.B. Barber
Mr. F. Beaumont
Mr. D.F. Bendall
Mr. C.V. Best
Mr. L. Burke
Mr. J.W. Byfield
Mr. D. Carr
Mr. J. Combes
Mr. D. Donald
Mr. P. Devoy
Mr. H. Donaldson
Mr. T. Evans
Mr. P.D. Foster
Mr. H.O. Fox
Mr. F.H. Freshwater
Mr. D. Gleeson
Mrs. G. Guley
Mr. J. Hammond
Mr. E. Harris
Mr. P. Hemsworth
Mr. H.L.R. Hudson
Mrs. V. Hutson
Mr. J.E. Janes
Mr. P. Kemp
Mr. W.G. Kew
Mr. R.G. Lamport
Mr. R.E.L. Laurent
Mr. T. Lewis
Mrs. H.K. Mawby
Mr. J.R. Miller
Mr. P.J. Moore

Mr. R.T. Allen
Mr. J. Beard
Mr. H. Belsey
Mr. D.W. Bertram
Mr. J. Blakemore
Mr. K.F. Butcher
Mr. A.R. Byrne
Mr. J. Clement
Mr. J. Cox
Mrs. G. Denton
Mr. J. Dewar
Mr. H.G. Dowsett
Mr. T.S. Evans
Mrs. E. Fox
Mr. P. Francis
Mr. L. Giebeler
Mr. G. Grimmel
Mr. C.G. Hamlyn
Mr. C. Hancock
Mr. G. Heather
Mr. A. Howard
Mr. A.J. Hunter
Mrs. N. Ives
Mr. W. Kay
Mr. A.E. Kennett
Mr. E.E. Lambarth
Mr. R. Launchbury
Mr. P.J. Lewis
Mr. D. Matthews
Mr. P.H.D. Merton
Mr. A.J. Moody
Mr. D.S. Morgan

Mr. L.S. Morton
Mrs. P.M. Murdoch
Mr. J. Murray
Mr. H. Owen
Mr. R.N. Paul
Mr. C.A. Pooley
Mr. J. Ratcliffe
Mr. R.C. Robinson
Mr. P.S. Sagoo
Mr. D.J. Shepperd
Mr. G.R. Smailles
Mr. T.E. Spoard
Mr. J.W. Tagg
Miss. F.M. Thomas
Mr. S.G. Thomas
Miss. D.E. Tiller
Mr. E. Walsh
Mr. P.J. Williams
Mr. A.L. Wyatt

Mr. J.T.E. Mumford
Mr. M.C. Murphy
Mr. T.F. Norton
Mr. H. Park
Mr. S. Phillips
Mr. D. Pugh
Mrs. A.K. Richardson
Mr. R.C. Rowe
Mr. D. Scully
Mr. J. Simmons
Mr. B.T.R. Smith
Mr. J. Swindells
Mr. J.T. Tennent
Mr. S. Thomas
Mr. W.H.A. Thomas
Mrs. B. Tripp
Mr. A. L. Warner
Mr. R.E. Williams
Mr. G. Webb

20 YEARS PLUS SERVICES

Mrs. J.M. Adams
Mrs. M. Allen
Mr. E. Ashmore
Mrs. C.E. Austin
Mr. R.H.E. Bailey
Mr. C. Barnes
Mrs. J. Barrett
Mrs. D.E. Begg
Mr. W.G. Bigby
Mr. H. Boocock
Mr. F.S. Bolger
Mr. F.A. Bowmer

Mr. R. Aldworth
Mr. R. Ashby
Mrs. I. Atack
Mr. E. Baggs
Mrs. C.E. Ball
Mr. R.W. Barnet
Mr. H.J. Baxter
Mr. M. Bharj
Mr. A.F. Birch
Mr. J.H. Booker
Mr. A. Borg
Mr. A. Bradley

Mr. L.J. Brill
Mr. F.G.H. Brooks
Mr. W.G. Burrows
Mr. K. Bryce
Mr. W.K. Cass
Mr. N. Clarke
Mr. T. Conroy
Mr. P. Cooke
Mrs. E. Coster
Mr. A. Darbon
Mrs. M. Davies
Mr. D.S. Donald
Mr. R.G. Dove
Mr. D.E. Duke
Mr. J.J. Dunn
Mr. S.P. Edmonds
Mr. R.H. Evans
Mrs. E. Fairbrass
Mrs. G. Fox
Mr. P. Francis
Mrs. M. Fulton
Mr. A.H. Gibbs
Miss. B.E. Gill
Mrs. A. Gilson
Mrs. R.M. Glenside
Mr. J. Gorodi
Mrs. K.A. Goultry
Mr. F.W. Harris
Mrs. M. Harris
Mr. M. Harrison
Mr. J. Hazell
Mr. R.A. Head
Mrs. H. Hill

Mr. M.J. Brooklyn
Mrs. D.N. Burrows
Mr. J. Brown
Mr. E. Byrne
Mr. G. Clark
Mrs. M. Collett
Mr. P.A. Cooke
Mr. K. Cooley
Mrs. R. Cusmans
Mr. D.G. Davies
Mr. W.T. Davies
Mr. G.L. Dove
Mr. D. Duckett
Mr. D. Duncan
Mrs. R.G. Dusgate
Mr. P. Ede
Mr. W.A. Evans
Mr. D. Forbes
Mrs. S. Fox
Mrs. K. Frowen
Mrs. T. Gannon
Mr. H.M.C. Gilbert
Mr. J.M. Gillespie
Mr. S.A.H. Girdler
Mrs. L. Godden
Mr. D.W. Gossedge
Mr. R.J. Harley
Mr. J. Harris
Mr. J. Harries
Mrs. J. Hawkesworth
Mrs. K. Hazell
Mrs. M. Herbert
Mr. J. Hind

Mrs. B.M. Hitchcock
Mrs. M. Hope
Mr. S.R. Inskip
Mr. A.W. Jones
Mr. D.L.V. Jones
Mrs. C. Kemp
Mr. B. Kinsman
Mr. G.B. Kirkum
Mrs. P. Lambarth
Mr. C.V.J. Last
Mr. C.W. Lewis
Mr. M. Lloyd
Mrs. M. Lowe
Mr. T.W. Mainwaring
Mr. J.W. Marshall
Mr. R.E. Matthews
Mr. H.E. McCoubrey
Mr. H. Meyerdierks
Mr. F. Millward
Mr. H. Monk
Mr. H.C. Morris
Mr. F.W. Mott
Mr. M. Murphy
Mr. C. Musgrove
Miss. V. Nogly
Mr. M. O'Halloran
Mr. C.V. Page
Mr. G.P. Peacock
Mr. D. Pennie
Mr. E.W. Phillips
Mr. G. Philpott
Mrs. H.E. Powell
Mr. G. Prentice

Mrs. P. Hood
Mr. W. Hutchinson
Mr. J.T. Jeffries
Mrs. B.F. Jones
Mr. R.J. Kellam
Mr. A.P. King
Mrs. M.L. Kirkbride
Mr. A.J. Knightley
Mrs. E.J. Lane
Mrs. D. Leakey
Mrs. E. Lichfield
Mrs. S. Long
Mr. F. Manley
Mrs. D. Marshall
Mr. A.F. Martin
Mrs. S. Mason
Mr. J. Meldrum
Mrs. D. Mills
Mr. J.T. Mitchell
Mrs. R. D. Mordaunt
Mr. A.M. Morrison
Mr. J. Munro
Mrs. B.J. Murray
Mr. P. Musiol
Mr. J. O'Connell
Mr. P. O'Shea
Mr. P. Parkes
Mrs. B. Pearce
Mrs. S. Penny
Mrs. M. Phillips
Mr. S. Poole
Mr. K.A. Powlesland
Mr. G. Pullen

Mrs. P. Randall	Mr. L. W. Ranson
Mr. M. Read	Mr. C.R. Robbins
Mr. E.C. Rose	Mrs. M. Rush
Mr. S. Rutter	Mr. J. W. Salter
Mr. R.A.G. Sarahs	Mrs. I. Saunders
Mr. L. Saunders	Mr. B.E.H. Shackle
Mrs. S. Sharpe	Mr. A.G. Shaw
Mrs. V.E.M. Shean	Mr. J. Sheeran
Mrs. Y. Shippen	Mr. G. Simpson
Mrs. S.A. Singer	Mr. E.W. Skinner
Mr. R. Smart	Mr. N. Smith
Mrs. S.F. Smith	Mr. T.V. Smith
Mrs. S.M. Snell	Mrs. L. Spearing
Mrs. N. Spowers	Mrs. F.M. Spuffard
Mr. T.C. Stephens	Mrs. L. Stephenson
Mrs. E.L. Stock	Mr. G. Stubley
Mrs. M. Tennant	Mrs. P. Thomas
Mrs. S. Thomas	Mr. A. Thompson
Mr. N. Thompson	Mr. R.K. Thompson
Mr. T. Tobin	Mr. P.D. Todd
Mr. W.H. Toull	Mr. S.C. Tubb
Mr. M. A. Tucker	Mrs. I. D. Turner
Mrs. L. Thornton-Lane	Mr. R.G. Vallintine
Mrs. P. Vidal	Mr. E. Wakeman
Mr. J.K. Walding	Mr. C.R. Walker
Mr. E. Walsh	Mrs. B. Walsh
Mrs. B. Ward	Mrs. M.E. Ward
Mr. R. Watson	Mr. C. Watters
Mr. D.J. Weaver	Mrs. I.L. Webb
Mr. A. Weeks	Mrs. G.S. Weeks
Mr. P.C. Welland	Mr. P.L. Welland
Mrs. M.A. Wharton	Mr. E. Whelehan
Mrs. C. Whitmarsh	Mrs. D.M. Williams

Mrs. W. Williams   Mrs. I.M. Wilson
Mr. J. Wilsher   Mr. L. Withey
Mr. G.M. Wood   Mr. J. Worley
Mr. D. Wright   Mrs. P. Wright
Mr. K.E. Yeo   Mr. H.A. Youssof

19 YEAR SPECIAL AWARDS
Presented 1992
AWARDED DUE TO PENDING REDUNDANCY

Mr. G. Allen   Mr. V. Cusmans
Mr. G. Hayden   Mr. D. Ironside
Mr. K. Harris   Mrs. D. Lumsden
Mr. F. Maloney   Mrs. B. Rickarby
Mr. M. Veness   Mrs. M. White
Mr. P.N. Wood

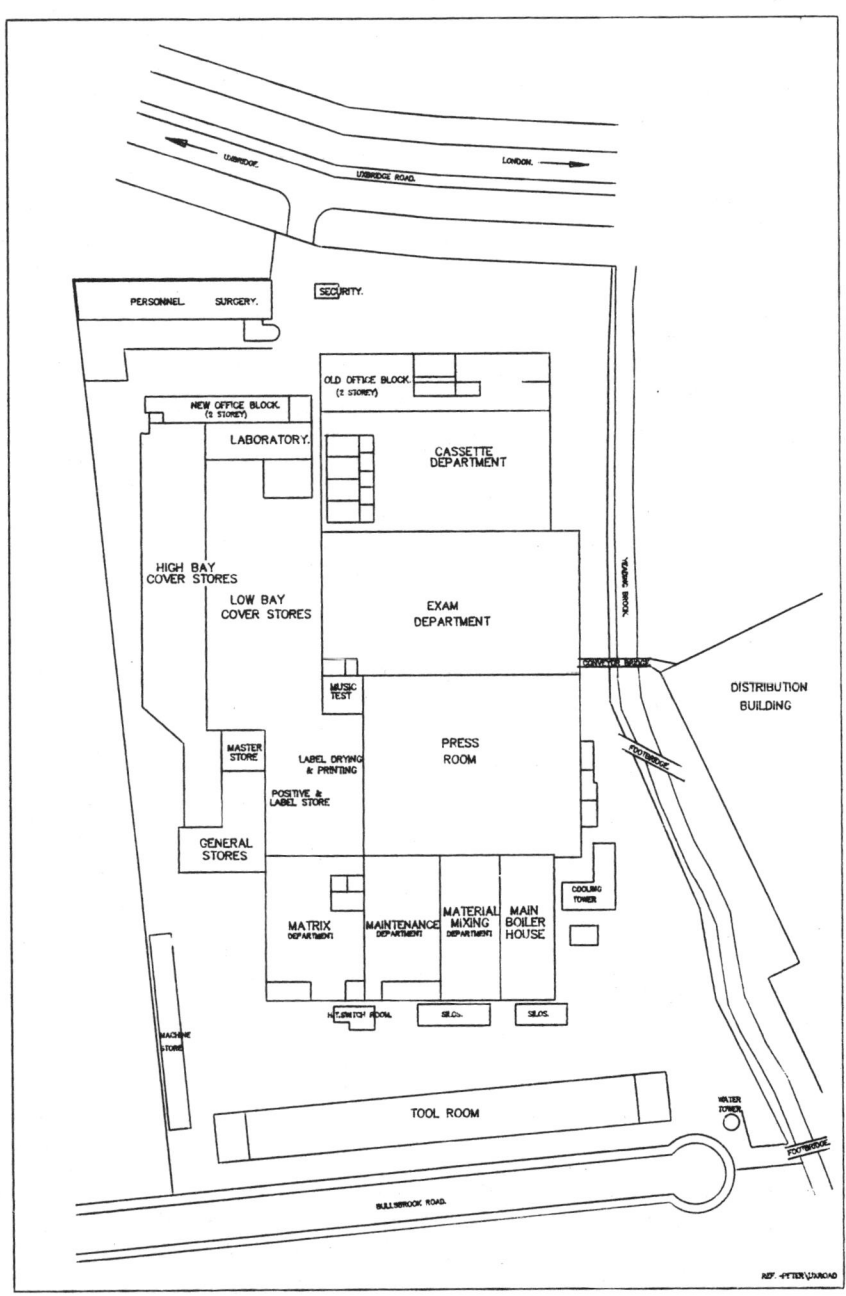

Plan of the Uxbridge Road manufacturing facility.